HEALING WATERS

Geographies of Health

Series Editors
Allison Williams, Associate Professor, School of Geography and Earth
Sciences, McMaster University, Canada
Susan Elliott, Dean of the Faculty of Social Sciences,
McMaster University, Canada

There is growing interest in the geographies of health and a continued interest in what has more traditionally been labeled medical geography. The traditional focus of 'medical geography' on areas such as disease ecology, health service provision and disease mapping (all of which continue to reflect a mainly quantitative approach to inquiry) has evolved to a focus on a broader, theoretically informed epistemology of health geographies in an expanded international reach. As a result, we now find this subdiscipline characterized by a strongly theoretically-informed research agenda, embracing a range of methods (quantitative; qualitative and the integration of the two) of inquiry concerned with questions of: risk; representation and meaning; inequality and power; culture and difference, among others. Health mapping and modeling, has simultaneously been strengthened by the technical advances made in multilevel modeling, advanced spatial analytic methods and GIS, while further engaging in questions related to health inequalities, population health and environmental degradation.

This series publishes superior quality research monographs and edited collections representing contemporary applications in the field; this encompasses original research as well as advances in methods, techniques and theories. The *Geographies of Health* series will capture the interest of a broad body of scholars, within the social sciences, the health sciences and beyond.

Also in the series

Towards Enabling Geographies:
'Disabled' Bodies and Minds in Society and Space
Edited by Vera Chouinard, Edward Hall, and Robert Wilton
ISBN 978 0 7546 7561 7

Geographies of Obesity: Environmental Understandings of the Obesity Epidemic
Edited by Jamie Pearce and Karen Witten
ISBN 978 0 7546 7619 5

There's No Place Like Home: Place and Care in an Aging Society
Christine Milligan
ISBN 978 0 7546 7423 8

Healing Waters
Therapeutic Landscapes in Historic and Contemporary Ireland

RONAN FOLEY
National University of Ireland, Maynooth, Ireland

Routledge
Taylor & Francis Group

LONDON AND NEW YORK

First published 2010 by Ashgate Publishing

2 Park Square, Milton Park, Abingdon, Oxon OX14 4RN
711 Third Avenue, New York, NY 10017, USA

Routledge is an imprint of the Taylor & Francis Group, an informa business

First issued in paperback 2016

British Library Cataloguing in Publication Data
Foley, Ronan.
    Healing waters : therapeutic landscapes in historic and
    contemporary Ireland. -- (Geographies of health)
    1. Hydrotherapy--Ireland. 2. Hydrotherapy--Ireland--
    History. 3. Therapeutics, Physiological--Ireland.
    4. Therapeutics, Physiological--Ireland--History.
    5. Health resorts--Ireland. 6. Health resorts--Ireland--
    History. 7. Health behavior--Ireland. 8. Health behavior--
    Ireland--History.
    I. Title II. Series
    615.8'53'09415-dc22

Library of Congress Cataloging-in-Publication Data
Foley, Ronan.
    Healing waters : therapeutic landscapes in historic and contemporary Ireland / by Ronan
Foley.
        p. cm. -- (Ashgate's geographies of health series)
        Includes bibliographical references and index.
        ISBN 978-0-7546-7652-2 (hardback) -- ISBN 978-0-7546-7653-9 (e-book)
    1. Landforms--Ireland. 2. Landforms--Northern Ireland. 3. Landscapes--
Ireland. 4. Ireland--Historical geography. I. Title.
    GB436.I73F65 2010
    304.209415--dc22

        2010004876

ISBN 13: 978-0-7546-7652-2 (hbk)
ISBN 13: 978-1-138-26026-9 (pbk)

Contents

List of Figures and Tables

Figures

Table

List of Boxes

Preface

Writing a book is often a personal journey and for me this is particularly germane. With a background initially in historical geography and later in GIS, my own personal development as a health geographer is a relatively late one. But having worked at doctoral level on access to services for informal carers in Southern England, my attention slowly shifted from an empirical and material interest in mapping and explanation, to a more nuanced interest in people and the spatial understanding of health behaviours. In considering the experience of carer's lives and the complex reasons that underpinned service utilizations and behaviours, I began to understand that there were complex negotiations, especially prominent in health and social care, between individual circumstances, experiences, feelings and responses and the constraints, wider spaces and cultural settings in which these operated.

As I began to teach courses in health geographies, initially at the University of Brighton and later at the National University of Ireland, Maynooth, that interest in explaining how people, place and health could be understood led naturally to an interest in 'the cultural turn' and its associated theoretical perspectives. That interest has led me to the study of the material elements of therapeutic places and the extent to which those places are culturally constructed around metaphors of health and water. However, some of my inbred empiricism (expressed by an interest in how the real world actually functions) never left me and drew me back to an interest in grounded therapeutic landscapes. But it is not an either/or perspective and I have become increasingly interested in taking a more relational approach which additionally explores experiences and inhabitation. This book is therefore a critical analysis of a set of healing places that are real and imagined, tangible and intangible, historic and contemporary and are shaped by a constant balancing act between inner meaning and outer context.

When asked if his work was inspired by other musicians, the great American guitarist, Leo Kottke once wrote, 'we all go through someone'. This is certainly true in my more humble case, given the breadth of the material I have drawn from across a range of different geographies (cultural, medical, economic and tourist) and wider disciplines (anthropology, folklore, medical and social history). Acknowledging that this book cannot possibly cover all of these literatures in detail, the citations and quotes used should point readers to the excellent work of other writers and researchers. Much of that wider research focuses primarily on a particular place or a specific time period, though a few take a broader perspective in looking at a particular type of place in a particular time period. However, there is limited cross-fertilization of either place or period. I feel that a book that attempts, even in a broad way, to look at different types of therapeutic landscapes at different times in different places, may extend the canon. In particular, there are connections, both empirical and theoretical, that link therapeutic places, and this is especially true of the watering-place in its multiple forms.

While watering-places have a global dimension, for practical purposes the book will focus on their development and cultural creation in one country, Ireland. The choice of Ireland is intended as a representative example, and the same approach and breadth of study could be applied to many other country settings as well. There are a number of contextual reasons for choosing Ireland, relating to culture, colonization and indigenous forms, which are explained more fully in the introductory chapter. As there are several thousand watering-places within Ireland, a representative sample has been chosen, but by the same token, many others have not been included. It is hoped that the approaches and sources used may encourage other researchers, both within Ireland and in other jurisdictions, to fill in those gaps and draw out deeper insights.

A wise man once advised John O'Donohue, the Irish philosopher and writer that; 'most research endeavours to establish a conclusion or reach verification that no-one can successfully criticize or undermine' (O'Donohue, 1997). O'Donohue's alternative to this defensive positionality was to acknowledge that you can't please everybody and think instead of new questions. In essence this is what the book will aim to do, drawing from older phenomenological traditions of person-place interactions to frame new thinking on the therapeutic landscape that begins to consider performance, inhabitation and living mobile narratives of place. These will be embedded within more formal themes which relate primarily to health and healing and its embodied, cultural, spatial and economic dimensions. All of those terms are relevant in exploring the role of psycho-cultural performances in understanding how people negotiate and interact with healing waters in place.

Throughout the book, Irish language names for both places and concepts are occasionally mentioned. In the book they will be referred to as the 'Irish' names and not, as perhaps might be better understood in the international market, as 'Gaelic'. Irish people rarely, if ever use the term Gaelic to describe a native word, preferring instead to either use the term, 'the Irish name', or if speaking the actual language, the term '*as Gaelige*' (from the Irish). So there is certain sensitivity to the word Gaelic, which the author wishes to acknowledge from the outset, while fully acknowledging the natural instinct of the non-Irish reader to look for, and use, that term.

Acknowledgements

No book is produced entirely alone and I would like to thank a large number of people for their help and support in the completion of this book. First and foremost, I would like to thank my colleagues in the Geography Department at the National University of Ireland for both friendship and encouragement over the past few years. In particular thanks are due to my Head of Department, Mark Boyle for allowing me the time and space to write and for regularly convincing me it was worthwhile. I would also like to thank those colleagues, Paddy Duffy, Rob Kitchin and especially Mary Gilmartin who all provided honest and extremely helpful feedback on individual chapters. Key support was also received from Jim Keenan, who drew me a lovely map and Mick Bolger, who got me a lovely laptop. I would also like to thank Mary Weld and Gay Murphy for their administrative support around finances and travel and general patience. Our growing band of health/medical geographers, Dennis Pringle, Fionnuala Ní Mhordha, Jan Rigby and Martin Charlton were also very supportive. Thanks also to other colleagues who chipped in with useful material and advice, especially Steve McCarron, Shelagh Waddington, Paul Gibson, Brendan Bartley, Conor McCafferty, Mary Kelly and Proinnsias Breathnach. David Meredith provided some very interesting perspectives and ideas, especially on the modern spa.

Thanks are also due to the National University of Ireland, Maynooth and especially the former Head of Geography, Jim Walsh, for allowing me to take a sabbatical year, during which the bulk of the writing took place. Without that time and space I would have been much delayed in submitting a manuscript. In addition, the advice received from Gillie Bolton at an NUIM sponsored writer's retreat in January 2008 was invaluable and I would like to thank her and the other participants for their suggestions and recommendations. Michelle McGuane in the Finance Office was also a star as were the library staff, especially Mary Antonesa. Outside of Geography, academic colleagues including Lawrence Taylor and Jacinta Prunty provided valuable material and advice. Finally thanks to large numbers of undergraduate students of health geography who inadvertently allowed me to try out some ideas on them over the past few years, especially the brave few who did projects on therapeutic landscapes.

I was fortunate enough to spend time in New Zealand while working on the book. I would especially like to thank colleagues at the School of Geography, Geology and Environmental Sciences at the University of Auckland, where I spent the winter of 2008–9 as a Research Visitor. Particular thanks to Robin Kearns for arranging the visit and for friendship and an amazing trip to the Hokianga and the Healing Waters of Ngawha. I would also like to thank Anne-Marie Simcock who set me up in the visitor's office with unbelievable efficiency and speed. My temporary room-mate, John Andrews was both a kind host and shared my interest in landscapes while I would like to thank the Head of School, Glenn McGregor for making me welcome and allowing me to skulk about for five months. Research students Abbey Wheeler, Tara Coleman and Christina Engler were also helpful and friendly. I would also like to thank Veronica Strang from Auckland's Anthropology Department for the invite to the conference, the nice brunch in Devonport and the sound advice ('structure, structure, structure') on writing a book.

One of the nice things about being a geographer in Ireland is that there is a strong disciplinary community across the country and I would also like to acknowledge support and useful comments at conferences (and in the occasional back-room of a hostelry), from colleagues at; TCD (Charlie Travis), UCD (Willie Nolan), DIT (Kevin Griffin), UCC (Willie Smyth and Dennis Linehan), NUIG (Marie Mahon and Andy Power) and Limerick IT (Frank Houghton). The other nice thing about being a health geographer is that there is an equally warm and welcoming global community who make the biennial Medical Geography Symposium an increasing pleasure. The friendship and support of colleagues at those conferences is especially valued and thanks to Anne-Cecile Hoyez, Thomas Kistemann, Charis Lengen, Geoff DeVerteuil, Christine Milligan, Sarah Atkinson, Damien Collins and Ross Barnett for their advice and comments over the years. In particular I would like to acknowledge the support of Allison Williams, who encouraged my book application and provided sound advice and support as commissioning editor. I would also like to thank Val Rose, Katy Low and Elaine Couper at Ashgate for their advice and prompt responses to numerous queries.

Given the breadth of the material in the book I was also very fortunate to have a number of external readers who all provided valuable and supportive advice. I would especially like to thank historians James Kelly (St. Patrick's Drumcondra) and Alasdair Durie (Stirling) and folklorist Stiofán Ó'Cadla (UCC), for reading the draft spa town, sea-bathing and holy-well chapters respectively. A mine of information and the creator of one of the finest websites around, Malcolm Shifrin, was also a helpful reader on Turkish baths. I have barely scratched the surface of Malcolm's encyclopaedic and meticulous material in this book. Melanie Smith from Corvinus University, Budapest and John Walton of Leeds Metropolitan University were also valuable contacts from the area of Tourism Geographies. Other Academic contacts outside of geography who were helpful were Angela Bourke of UCD and Irene Furlong, formerly of NUI Maynooth, who both acted as valuable sounding-posts.

Another great pleasure in the work was actually getting out in the field, both literally and metaphorically. As a three year piece of detective work that took me from muddy mountain sides to plush spa interiors, I encountered many helpful people in many different settings. Apologies if those people are not all listed here, but I did appreciate the material, documents, direction and most of all stories, that I received. From the local history offices, archives and journals across the island of Ireland I would like to thank Peter Beirne, Ciaran Parker, Vincent Allen, Susan

Garrett, Mary Bradley, Gemma Ward, Brigid Loughlin, Loreto Guinan, Mario Corrigan, Caroline Martin and Noel Ross. Given I have sites from 22 different counties, there are many more staff at archives, libraries and local studies centres who I should thank and do anonymously. In addition, local experts, writers and overseers at different sites, both ancient and modern, were a mine of information. Betty Newman-Maguire was a great source and a kind hostess (as well as being a superb sculptress) at St. Kieran's, while Marie Heavey-Bartley, Tim O'Brien, Stephen Talbot, Oliver McCormack, Sr. Mary Minihan and Maudy Guthrie also gave me excellent local information. In the contemporary settings of the modern spa and seaweed baths, I would like to thank interviewees including; Alison Bell, Suzanne Mulraney, Patrick Sibirski, Liam Griffin, Liam Griffin Jnr, Eileen Doyle, Michelle Moloney, Adriaan Bartels, Mohommed Abou Saleh, Alyison Smith, Declan Fagan, Ciara Wilson and Dr. Anu Gopautiah. The wonderful tradition of the seaweed bath is being kept alive by Seamus and Claire Mulvihill at Ballybunion, Edward Kilcullen at Enniscrone, Neil Walton and his family at Strandhill and Declan Devine at Newcastle. All of them made me welcome and even made me partake of the *fucus serrata*. I am also grateful to the staff at the National Folklore Collection UCD for permission to reproduce material. One of the other great learning experiences for me in researching the book was the joy of talking to people at all of the sites I visited. Here I very much depended on the 'conversational kindness of strangers' and heard many wonderful stories of cures, practices, rituals and ongoing inhabitations of places with therapeutic power. These were to be found all over Ireland but also in New Zealand where meeting people like Mike 'the Maori' O'Donnell at Paeroa, marked a focal point along a journey which continues on, in and towards water.

Finally on a personal level, my friends and family are heartily sick of yet another 'interesting' fact about watering-places. I promise I will stop after this, but thank especially my parents Conor and Nell Foley and siblings Deirdre, Emer, Connell and Cliona, all of whom have some place in the book in roundabout ways. The opinions of a journalist and development policy expert were especially helpful as external readers. I would also like to thank other good friends who have patiently listened over the past few years including; Peter, Laura, Dee, Polly, Gisele, Eoin, Mary G (again!), Mary L, Angela, Eamon, Pauly and Sharonne, Mick, Barry and Kirsty. Special thanks to Marita for helping me believe in myself and to the members of the culture club/book group, Theresa, Aidan, Jenny, Elspeth and Emily who have promised never to make this 'book of the month'.

Ronan Foley
Auckland, Lucca and Celbridge

Dedicated to my paternal grandmother, May Horgan, one of the first women in Ireland to own a car, who drove my father to the beach at Ardmore, setting in train a genetic love of the seaside.

Chapter 1
Introduction

Water, Health and Place

As a child growing up in land-locked Laois, far from the ocean, the highlight of the year was the family trip to the seaside at Kilkee on the West Coast of Ireland. Like most families, we held an in-car competition as to who first saw the sea. My sisters always won, yet the first sight of the Atlantic brought with it a burst of excited feelings linked to the water and what we would do in the place. As an adult living by the banks of the Severn in Worcestershire, England, a typical Sunday walk along the river passed a small colony of shacks. These were informal and ephemeral homes to escaping Black Country factory workers, for whom Bewdley was the nearest place they could get their fix of nature and water. In both settings the power of water, maritime or fluvial, and place, constructed or temporary, combined to provide a simple, but affective sense of well-being for individuals, families and communities.

We all know that water is essential to life, we need to consume at least two litres a day to maintain that life while access to clean healthy water continues to be one of the most significant barriers to global health (Kingsley et al., 2009). Water has many other roles in everyday life and is endowed with particular qualities. This book focuses on one of those qualities, namely the ascribed healing power of water and its material expression in place. Human associations with water as physical healer are deeply embedded at a global level (Strang, 2004; Gerten, 2008). In particular these associations are expressed through and in geographical places, with springs, pools, rivers, lakes and the sea all associated with health-giving powers and constituting what may be termed therapeutic landscapes. Such landscapes are defined by Kearns and Gesler as; 'places that have achieved lasting reputations for providing physical, mental and spiritual healing' (1998, 8). The essential function of water as a life-preserving lubricant within the human body is also expressed at a more spiritual and elemental level in a wide number of religious and other-belief systems (Shaw and Francis, 2008). In such systems, rivers like the Ganga or lakes like Superior are believed to be the homes of the gods and as such, the guardians of life. In addition, special reverence is often held for those places where water courses begin, such as the source of the Indus at Mount Kailash in Tibet (Shackley, 2001).

Water's ability to act as a metaphor for life, and by extension, health, is endless. In its mobile flow through the landscape, the course of a river is representative of the sinuous life-course experience of human health. In its often still presence as lake or pond, the therapeutic and calming natures of water are regularly invoked

(Conradson, 2005a). As raging torrent, or wild-waved sea, the volatile and ambiguous nature of water is visible as an element of contestable health, capable of both good and harm (Corbin, 1994). In such meanings, 'water has a nearly unlimited ability to carry metaphors' (Illich, 1986). These are often expressed in homologous form such as immersions in 'bodies of water' that emphasize embodied and emotional connections between human and aquatic environments (Strang, 2004; Shaw and Francis, 2008). The body, itself an average of 60% water, can also be seen as a 'permeable spatial form', through which water is absorbed, stored, circulated and expelled (Curtis et al., 2009). Water's symbolic and spiritual role within healing and well-being is also reflected in its utilization in a range of complementary, alternative and holistic practices across time and space (Williams, 1998; Heller et al., 2005). These reflect non-western and global understandings of health wherein metaphors of integration, balance and diversity are central tenets (Bergholdt, 2008).

There are a number of visible place expressions of water, with specific connections to health and healing that include springs, wells, seas, baths and spas. It is from such places within Ireland that this book draws its own inspiration and focus. Both natural and constructed, all of the forms, though often treated similarly, are of different provenance and meaning. The natural spring can be both emergent source and simple pool, while wells may be the pools which form at springs or become 'domesticated' through human agency. Sea-water helped create the coastal bathing resort and is used in modern practices of thalassotherapy as a literal 'health from the sea'. Baths are constructed by humans in many forms for hygienic, spiritual, pleasurable and healing (balneotherapy) purposes. The spa, defined as; 'a curative mineral spring, or a place associated with the same', is where 'water-health' becomes most explicit (Pearsall, 2001). It has been suggested that the Walloon word *espa* (meaning fountain) became associated with the eponymous town in Belgium in the twelfth century (van Tubergen and van der Linden, 2002). The Romans referred to their healing springs as spa, suggested as an abbreviation of the phrase, *sanitas per acqua* or 'health from water' (Jackson, 1990). In employing Kearns and Gesler's notion of therapeutic landscapes, the ways in which these differing forms of watering-place and their curative reputations are produced, sustained and contested lies at the heart of this work.

While healing watering-places have a global dimension, there are a number of contextual reasons for choosing Ireland as the setting for a book. Celtic cosmologies around nature and landscape were predicated on the spiritual and physical healing dimensions of place that reflected global indigenous cultural narratives. Trees and stones were part of that narrative but healing was also intricately connected with seas, lakes, rivers and springs (Rattue, 1995; Jones and Cloke, 2002). This focus on native and indigenous health may well resonate in other societies and settings. Secondly, Ireland's complex historical position as both part of the UK, but also its closest and most volatile colony, provides a rich colonial and post-colonial setting within which to examine healing waters (Said, 1985). The social and cultural production of watering place forms, such as spa towns and sea-bathing resorts, offers rich potential for similar analyses in other colonial settings (Breathnach, 2004; Kavita, 2002; Urry, 2002). Thirdly, there are a number of examples in Ireland which act as markers for the more globalized production and reproduction of healthy places (Hoyez, 2007b). Examples include the Turkish bath phenomenon of the late nineteenth century along with modern spa and wellness settings that represent a form of globalizing therapeutic landscape (Williams, 2007; Shifrin, 2009). In addition, there are some relatively unique Irish phenomena such as the sweat-house and seaweed bath; country-specific expressions of healing waters which may also inspire comparable work within other jurisdictions. Finally, a deeply embedded and common expression in Irish vernacular culture, 'The Cure', referring to a power certain people and places have, attests to deep phenomenological and cultural connections with the therapeutic powers of nature, in which water remains prominent.

In setting out the book, this chapter briefly introduces the theoretical ideas that underpin a therapeutic landscapes approach. Five different place-studies then form individual chapters that investigate a range of such landscapes, all broadly categorized as 'watering-places'. The specific settings include holy wells, spa towns, Turkish baths and sweat-houses, sea bathing resorts and the modern spa (see Figure 1.1). Each study investigates a specific water-based therapeutic landscape to study how that setting reflects a range of performances of health in historic and contemporary place. They also consider the natural/built, symbolic and individual/ social environments within which such enactments and experiences take place (Gesler, 1992; Williams, 1999b). A final short summary chapter pulls together connections across the different watering-places and identifies some cross-cutting themes for future research.

Therapeutic Landscapes

> Interactions with water take place within a cultural landscape which is the product of specific social, spatial, economic and political arrangements, cosmological and religious beliefs, knowledges and material culture, as well as ecological constraints and opportunities ... Cultural landscapes contain deep historical roots. (Strang, 2004, 5)

In the early 1990s, and marking a shift from the earlier medical geography focus on epidemiology and health care planning, culture became more deeply embedded in explanations of health in place (Kearns and Gesler, 1998; Williams, 1999a; Kearns and Moon, 2002). In beginning to link cultural and health geographies through this period, the term landscape was heavily used with a particular focus on its cultural and phenomenological dimensions (Wylie, 2007). In both traditions, a notion of landscapes as symbolic, 'as expressions of cultural values, social behaviour and individual actions worked upon particular localities over a span of time' (Meinig, 1979), remained central. Such visions were attractive to health geographers who identified ways in which landscape, as a metaphor for place, could be reutilized as a central plank in new formations of the subject.

> In these various guises the idea of 'landscape' has sought to convey many different meanings. For some, it is analogous to literally defined localities. For others it is a metaphor for the complex layerings of history, social structure and built environment that converge in particular places. Though its differing meanings suggest a degree of pluralism which sometimes borders on the chaotic, there is also a sense in which, notwithstanding its internal inconsistency, it remains the term that most clearly embodies the tropes of place and health that were expected to be the hallmarks of a new geography of health. (Kearns and Moon, 2002, 611)

The attentions and interests of health geographers coalesced around a number of such landscapes (Gesler and Kearns, 2002). The term was applied to important investigations of health inequalities, drawn together under such banners as landscapes of exclusion, fear, risk and consumption (Curtis, 2004; Gatrell and Elliott, 2009). However, the emergent sub-theme most relevant to this work was the notion of the therapeutic landscape (Gesler, 1992; Williams, 1999c).

Initially developed by health geographer Wil Gesler and focused on 'traditional' settings, such as Epidauros and Bath, that work identified how symbolic and reputational narratives of health, as well as their material expression, were centrally involved in place creation and sustenance (Gesler, 1992; 1993; 1998; Williams, 1999a). The healing effects of place were also explored through natural and built environments as well as in more everyday therapeutic encounters experienced in place (Palka, 1999; Thurber and Malinowksi, 1999). Here the physical surroundings, whether majestic (in wilderness settings like Denali National Park) or domestic (in home, park, wood or garden), were connectively linked, in health terms, to earlier geographical explorations of 'senses of place' and perceived positive phenomenological connections between wellness and place (Tuan, 1974; Relph, 1976). In such natural and semi-natural settings, a mix of wood, stone, sky and water were linked to an affective response from humans, often expressed in spiritual terms as a manifestation of the sacred (Eliade, 1961; Graber, 1976) The therapeutic landscape concept was extended to a focus on design, allowing geographers to explore how specific built environments could be positively shaped for the improvement of health based on aspect, materials and affective elements of light, colour, space and, not-uncommonly, water (Kearns and Barnett, 1999; Curtis, 2004).

Much of the initial research on 'traditional' therapeutic landscapes focused on places associated with water (Smyth, 2005). Gesler's studies of spa towns and pilgrimages sites such as Lourdes and Bath were particularly relevant (Gesler, 1996 and 1998). At Lourdes, the heavily commodified pilgrimage site in Southern France, the symbolic importance of water was central to narratives of reputational healing. In the stories of the curative power of its holy waters, place, setting and affect were combined, along with a sense of what Turner termed communitas, as expressions of communal faith healing (Taylor, 1995). At the same time Lourdes was identified as a highly commodified setting where the holy water became, alongside other icons and souvenirs, an essentially portable product (Gesler, 1996). These juxtapositions of health, belief and profit were common to many of the watering-places considered in this book. Water was also central to the historical reputation of Bath as a healing place (Gesler, 1998). From its original Pagan/Christian wells, to the Roman spa and its later Georgian manifestation, the hot springs at the site provided a physical and discursive connection to place and healing, evident in contemporary forms such as the Thermae Bath Spa (Gesler, 1998; Sharpe, 2006).

These connective uses of the health histories of place were also to be found in studies of spa towns in other settings, though relatively few were by health geographers. Geores' work on Hot Springs in North Dakota identified the importance of metaphoric health narratives in the sustenance of place reputation from initial American Indian spiritual cure stories to its nineteenth-century commercial marketing and beyond (Geores, 1998). In the metaphor of 'health=hot springs', the selling of health through the invocation of the name reflected similar discursive uses of the word spa in the contemporary world. More broadly, while individual spas in different countries had their medical and social histories extensively reviewed, critical place-based health perspectives remained underdeveloped (Rockel, 1986; Brockliss, 1990; Hembry, 1990; Durie, 2003a). Valenza's survey of Texan hot springs touched on the importance of embodied experiences of health in place, though this was a rare example of such an approach (Valenza, 2000). It was as if the early explorations of the spa town or pilgrimage site as exemplary therapeutic landscapes was taken as an imperative to develop the subject into any other site or setting, leaving those foundational sites surprisingly under-explored, despite the enormous global range of watering-places.

In considering 'the continuing maturation' of the therapeutic landscapes concept, two strands seem particularly relevant; the contestation of therapeutic assumptions and the formative role of experiential encounters in place (Hoey, 2007; Williams, 2007). A more critical approach to the assumed health benefits of encounters within therapeutic landscapes represented an important shift in thinking. In their work on the beach in New Zealand, Collins and Kearns identify the pharmaconial nature of these settings in health terms, through their potential ability to both cure (via connections with rest, leisure and exercise) and kill (from melanoma and drowning) (Collins and Kearns, 2007). Davidson and Parr's work on phobic spaces extended the notion that one's experience of place could also be negative/excluding and that the need for 'safe havens', echoing earlier mental health research by Pinfold, was an important part of person-place interaction (Pinfold, 2000; Davidson and Parr, 2007). Monolithic assumptions around socialization, place and health needed to be recast as individual imaginative negotiations where even being in certain places could be anti-therapeutic. In addition, therapeutic sites associated with water such as bathhouses, were re-assessed in terms of risky health behaviours, especially in relation to the risk of HIV/Aids infection at gay bathhouses; a setting specifically associated with the start of that global pandemic (Andrews and Holmes, 2007). Such liminal activities and health-risk associations have lengthy behavioural histories, expressed in a variety of social and sexual forms, at other watering-places such as public baths, springs, wells and the seaside (Shields, 1991; Corbin, 1994; Lenček and Bosker, 1998). These links to health behaviours emphasize the importance of individual inhabitation and performance in therapeutic outcomes, and these inherent contestations of health are revisited in the Irish setting (Davidson and Milligan, 2004).

As a closely-linked variant, such settings can also be seen as landscapes of therapeutic potential, shaped by personal and communal performances and wider structural forces. In challenging an assumption that places are inherently therapeutic, the importance of an individual's own experience and response may create a positive or even negative effect. Using wider metaphors of retreat, escape and stillness, Conradson identifies the significance of the 'therapeutic landscape experience' as a relational outcome (Conradson, 2005b). In these nuanced interpretations,

> particular landscapes are found to be not intrinsically healthy or unhealthy; rather they may be used, experienced and perceived differently by different people ... In general terms, a therapeutic landscape experience might then be understood – from a human point of view – as a positive physiological and psychological outcome deriving from a person's imbrications within a particular socio-natural material setting. (Conradson, 2005b, 339)

In that experiential encounter there are ties to Andrews, Sudwell and Sparks' (2005) recent observations about the need to also consider the importance of imagined and affective therapeutic landscapes as much as material, physical ones.

While recent research on therapeutic landscapes has begun to focus on the experiential, and acknowledges the potential contradictory experiences that individuals and communities may have in place, these must be seen in the light of the wider contexts of the place. Additional recent work by health geographers has begun to revisit the connections between a sense of place and health, with a focus on both stable everyday as well as displaced settings (DeMiglio and Williams, 2008; Eyles and Williams, 2008). The continuing importance of culture, society and economy emphasizes how place remains active in shaping health and how health is, in turn, affected by, through and in place. While there are many 'internal inconsistencies' within the study of therapeutic landscape, it is precisely those contradictory narratives, identities and performances which continue to make them rich settings for geographical enquiry (Kearns and Barnett, 1999). In beginning to think around contemporary re-workings of the therapeutic landscape, an early definition of Gesler's suggests twin components; 'Inner/meaning (including the natural setting, the built environment, sense of place, symbolic landscapes, and everyday activities) and Outer/societal context (including beliefs and philosophies, social relations and/or inequalities, and territoriality)' (Gesler, 1993, 173). In considering the twin roles of setting and context, therapeutic landscapes can, in part, be studied as material and empirical settings. Yet there is a deeper richness in examining those twin roles as enacted spaces of negotiations between inner and outer meanings.

Performances of Health

In considering how different watering-places are created and sustained, the notion of performance is central. The recent focus within therapeutic landscapes research around inhabitation and experience draws in part from wider cultural theory and what has been broadly termed, the 'performative turn' (Wylie, 2007). A backlash against overly representational approaches, the development of non-representational theory was linked to a deeper interest in how spaces were lived and inhabited, or the 'texture of space and place rather than its textual representation' (Hubbard, 2005; Thrift, 2008). Some of this research drew on a phenomenology where landscape was, 'conceptualized in terms of active, embodied and dynamic relations between people and land, between culture and nature more generally' (Wylie, 2007, 143). In addition, terms like 'dwelling' were used to emphasize these relationships in a performative sense through repeated encounters and persistent patterns of flow (Jones and Cloke, 2002). Some of these 'dwellings' were also considered to have a spiritual expression which connected to landscape qualities, well-being and health (Smith, 2005; Wylie, 2007). Within the performative turn, 'feelings' were also identified as a significant aspect of that deeper connection between inhabitation, action and place. The notion that self-landscape encounters and experiences could be expressed through moods, senses, feelings and emotions also related back to wider phenomenological and more-then-representational considerations (Lorimer, 2005; Tolia-Kelly, 2006). The emergence of emotional geographies and in particular the near-ubiquitous term, affect, within that research, further emphasized these theoretical connections between place and mood as an underused aspect of therapeutic landscapes (Anderson and Smith, 2001; Davidson and Milligan, 2004).

In considering performance more fully, 'dwellings' and 'feelings' were also expressed in 'doings', whereby individuals managed their own enactments and inhabitations in space (Merleau-Ponty, 1962; Thrift, 2004). These enactments had a particular relationship to space and place, in shaping where and how performances were produced, created, managed and understood. Less fully applied to the health/place arena than they might be, the potential of applying a set of 'relational performance' themes to an exploration of healing at the watering-place is both a challenge and an opportunity (Wylie, 2007). Within a wider field of health identities and beliefs, such actions could be formal, ritualized and normative but also resistive, transgressive, liminal and contested. In considering the creation of healthy spaces, the wider powers and structures involved in the creation of such places could be subtly altered to take a personal performative agency into account (Lorimer, 2005). In furthering a connection in terms of embodied practice; 'non-representational geographies conceptualize the body as sensuous, sensitive, agentive and expressive in relation to the world, knowing and innovating amongst contexts and representations that become reconfigured in practice'. (Crouch, 2000, cited in Jones and Cloke, 2002, 8)

For health, this connection to how it is created in place via complex active and lived performances can also be seen in terms of what Nash, in her valuable comparison with artistic forms such as theatre and dance, describes as, 'micro-geographies of habitual practices' (Nash, 2000). While the representative can remain important in terms of the wider contexts around which place is constructed, its internal meanings and identities are an expression of practice within it and the reputational identities that in turn creates. This also speaks to a biopsychosocial model of health rather than a biomedical one (Blaxter, 2004; Heller et al., 2005). In developing these ideas within watering-places, performances of health can be explored under four main headings namely; bodies, cultures, spaces and economies.

Bodily Performances of Health

In beginning to think about practices and performances of health in therapeutic landscapes, the body lies at the centre of such enactments and is the locus of health outcomes. Such outcomes can range from verifiable medical efficacy to the more common imaginative therapeutic experiences of the placebo, the faith cure or the rest cure. It must be acknowledged that from biomedical and rational perspectives, the medical relevance of different forms of watering-place like wells, baths and spas are at best irrelevant, at worst dangerous (Porter, 1990). These responses are expressed in a variety of contestations that reflect the protection of a status quo of power, control and ownership, but that also reflect wider public opinion which values scientific medicine above all other forms. Nevertheless, the book considers a range of embodied water cures, treatments, associated health services and lay health beliefs in Ireland, while framing these against wider contestations evident in medical histories and in the writings of cultural and medical/health geographers.

In considering bodily performances of health, a debt is owed to feminist geographies, where embodied concerns with the body as a space of meaning and expression were first discussed (Rose, 1993). In identifying the body as a site of experience, performance and feeling, bodies themselves could be transformed into landscapes of healing, expressed in inscription on the body as well as imagined from within it (Parr, 2002; English, Wilson and Keller-Oloman, 2008). Thus embodiment is essential to any consideration of water and health, being the site of medical exploration and intervention but also where one's own feelings of healthiness and well-being dwell. In any patient-healer interaction, the patient is less likely to describe their exact symptoms as express that they 'do not feel well' or point to where it hurts; while the journey from that feeling to a successful health outcome requires a narrative engagement between the patient's body and the healer. That narrative between patient and healer also raises embodied concerns around control and power, and how healthy and unhealthy bodies are managed and controlled, a concept central to the work of Foucault (Philo, 2000; Kearns, 2007).

While the term affect is hard to clarify and is often used as a synonym for emotion, it has a wider embodied meaning, linked to senses and moods, which aligns well with the therapeutic experiences humans have in place (Philo and Parr, 2003; Tolia-Kelly, 2006). Relationships between the senses and place are simultaneously invoked in their intensities of response, given that 'affect denoted the shifting mood, tenor, colour and intensity of places and situations' (Wylie, 2007, 214). In therapeutic landscapes terms, that sense of place can just as readily be rendered as 'places of the senses', wherein:

> ... there is a strong interplay between emotions and locations of healing. Such an interplay speaks to the existence of what Davidson and Milligan (2004, 52) term an 'emotio-spatial hermeneutic' – 'emotions become understandable only in the context of particular places. Likewise, place must be *felt* to make sense' (original emphasis) ... places become landscapes of healing due to the strong emotional ties embedded within everyday and extraordinary locations. (English, Wilson and Keller-Oloman, 2008, 76)

The ways in which individuals and cultures draw spiritual and psychological healing from particular settings and places like holy wells, rivers, lakes or the seaside also link to what Williams refers to as 'landscapes of the mind', especially where that has an imaginative element in how people interpret space and place and process that for therapeutic purposes (Williams, 1998; Eyles and Williams, 2008).

Cultural Performances of Health

Given that health and healing is enacted in real places by real people, such enactments can be framed as cultural performances of health. In considering culture, two main strands of theoretical thinking inform the book. The first reflects Jackson's (1989) notions of maps of meaning, where individual inhabitation and agentive power shape the felt and symbolic potential of places. The second strand draws on an understanding that culture is also shaped by social relations and the wider structural agency of politics, society and economics (Kearns and Gesler, 1998; Mitchell, 2000). While these two definitions may seem contradictory, the ways in which these contradictions are intertwined and negotiated significantly informs this work. From a health perspective, lay experiences, beliefs and perceptions at the 'watering-place' are therefore, informed and shaped in turn by wider cultural models of health and well-being (Porter, 1992). They also echo a shifting moral geography in place, reflected historically in how faith and spirituality informed understandings of health, where sickness and death were not physiological abnormalities, but acts of a sometimes vengeful God (Porter, 1999). While the current hegemony of a biomedical model of health is evident, these earlier relationships between mind, body and spirit are revisited in the increasing popularity of complementary and alternative medicines (CAM) in contemporary society (Heller, 2005; Williams, 2007). These more spiritual and culturally-formed relationships also reflect the historical construction of healing places and connect historic and contemporary understandings of therapeutic landscapes.

Healing waters are the product of a range of cultural narratives and performances, from the religio-magical to the pseudo-scientific, wherein healing is expressed through words and feelings as much as in physiological outcomes. Ranging from the utterly dismissive to the utterly credulous, it is in the stories of health and the places where that health is produced that the watering-place assumes meaning and identity. The milieu of the watering-place is also shaped by wider inhabitations tied to politics, power and identity. The ways in which health is experienced and managed reflects in part, how individuals and communities enact and contest specific health beliefs and expectations. In turn, the cultural and symbolic importance of those performances can tell us much, as in Ireland, about the agentive social processes that form them and the wider moral geographies implicit in their sustenance or transformation. In considering the extent to which performances of health are culturally created, Kleinman's much-cited commentary notes that; 'Healing occurs along a symbolic pathway (continuum) of words, feelings, values, expectations, beliefs and the like which connect events and forms with affective and physiological processes' (Kleinman, 1973, 210). While there is an embodied dimension, it is through cultural expressions of values, expectations and beliefs that narrative and healing 'pathways' are produced and, just as importantly, reproduced (Kearns and Barnett, 1999; Gesler, 2003). Hubbard suggests that this reconsideration of repeated performance, expressed in images, symbols and metaphors, is a further by-product of the performative turn (Hubbard, 2005).

A second theme running through the book focuses on such cultural performances of health. In Ireland these were enacted in vernacular settings like holy wells and sweat-houses, as well as the more colonial spaces of the spa or sea-resort. They also reflected different cultures of health practice, between folk and scientific medicine. In addition there were quite conflicting expressions of curative and social identities at all of the watering-places. In particular the use of such settings for leisure and pleasure, as well as health and healing, were significant cultural acts. In the often liminal and profane behaviours at the spa or well, cultural prohibitions and allowances also shaped the relative positions of health and culture in place. In considering these juxtapositions, a wider understanding of cultural performance, wherein inhabitations and experiences were expressed through rituals, class, colonial identity and 'native' behaviours, must also be deployed.

Spatial Performances of Health

A third theme takes a more specifically geographical approach, focusing on the spaces of health performance at the watering-place. This strand explores how place characteristics, topography and spatial networks all affected how watering-places were developed, maintained and altered across the island of Ireland. In looking at how the places were used, inhabited and experienced, their locations and wider issues of exclusion, class, gender and colonialism can also be invoked in better understanding the place of the patient/user of the therapeutic landscape. The potential to take innovative approaches to linking the geography of the place with the affective responses of the people in those places is also explored. While there are methodological difficulties in examining these more felt aspects of inhabitation, especially in historical settings, those therapeutic place dimensions remain an important structural element.

While embodiment and culture shaped how health was experienced and expressed, performances of health were enacted in tangible material spaces. In transient and repeatedly performed interactions of people and places, spatial performances of health were significant in shaping the therapeutic landscape (Hetherington and Law, 2000). The fundamental understanding of the therapeutic meaning and power in place was expressed in Ireland through specific health-place associations of rest and retreat to places where the natural and constructed surroundings were designed to cure. The landscapes were affectively set within woodland, rolling hill, lake or sea-shore or within intense indoor spaces of heat, steam and bath (Dunkley, 2009). In such settings, the patient was removed from their normal stressful daily setting and taken to spaces where they could relax and take the time to make themselves better. In these phenomenological connections, the place was capable of inducing a multiple response from the people inhabiting it, even if the full therapeutic benefit of that engagement could never be guaranteed (Conradson, 2005a). What mattered was that the place encouraged the possibility of a therapeutic experience due to its inherent qualities. Specific local topographies of lake, stream or valley, meant that health-place interactions were often unique and individual, though wider communal performances were also shaped in/by such settings.

Spatial performances of health were also mobile. Ephemerality and permanence were both visible in the spatial enactments of healing in place. As Conradson (2007) points out, the opposite of mobility, its balancing 'other', is found in expressions of stillness. For a geographer interested in these juxtapositions, where speed and stillness intertwine and are expressed in mobilities and non-mobilities, the relational position of the watering place can be usefully studied across time and space. While they were often short-lived and ephemeral, their health/place identities contained lingering, and at times, permanent components. Epidauros, as a historic site with a narrative therapeutic power sustained in contemporary commodification, represented such an identity (Gesler, 1993) In addition, these interactions were themselves expressed in a spatially performative sense, through access and utilization by a range of different clienteles coming from a range of different catchments. Yet even these were mobile over time as colonial rule gave way to independence and the spaces of contact between the colonial and native became more hybrid and fluid. While many watering-places might now be considered 'special', as opposed to 'everyday' places, the sheer variety and number of such settings meant that they had everyday and heterogeneous clienteles (Milligan, 2007).

Economic Performances of Health

As a final theme, a focus on economic performances of health is one which merits deeper attention in settings such as spa towns, sea-bathing resorts and modern spas, all of which have substantive literatures in medical, historical and tourism geographies (Porter, 1990; Walton, 2000; Urry, 2002). The fact that the majority of the Irish therapeutic landscapes were connected to the selling and commodification of the watering-place 'cure' emphasizes the importance of that economic dimension (Doel and Segrott, 2003). Many watering-places were developed as explicitly commercial ventures, with entrepreneurship and the role of risk being central to the production of place (Towner, 1996; Aitchison, Macleod and Shaw, 2000). While many were successful and brought wealth to the entrepreneurs as well as wider populations, watering-place histories were also littered with narratives of commercial failure or short-lived fame (Hembry, 1997).

   While tourism and leisure research had an implicit rather than explicit interest in health and healing, newer work provides valuable evidence identifying its importance in the creation and maintenance of the watering-place (Connell, 2006; Smith and Puczko, 2009). While the consumption of the waters was generally free, this was set within spaces shaped by a wider health consumption (Kearns and Barnett, 1999). The roles of investment, capital, fashion and marketing are crucial in understanding the development and sustainability of the places studied in the book and indeed more widely in how health is commodified (Smith and Puczko, 2009). As water was at the heart of place production, it was central to curative reputations, both physical and spiritual. As the product around which place was built, water was explicitly linked to its sacred, physical and chemical properties. These intrinsic qualities were then made explicit in their commodified forms and practices through metaphors such as 'healing waters' and even 'fountains of youth' (Valenza, 2000).

   In addition a consideration of an economic ownership of health provides a pathway into deeper considerations of structure, agency and power relations as performed in these sites. The extent to which individual/communal readings and 'livings' of health and well-being conflicted and conflict with the health visions of wider structural agencies of church, state, market and medical profession affected that performance. All of the watering-places were owned and managed in legal/financial terms, yet the ownership of health-place was also shaped by the extent to which individuals and communities owned their own health-practices. The potential to link concerns with individual therapeutic performances and experiences against the deeper formational forces shaping a particular expression of health in place, suggest that ownership was a mobile and negotiated process with strong economic foundations. For both producers and consumers this notion of 'place experience' is also one which resonates in tourism geographies, where notional 'atmospheres' of place are felt and enacted, yet also commercially structured (Smith and Puczko, 2009).

Re-placing Traditional Therapeutic Landscapes

While there is a new and valuable focus in therapeutic landscapes research on everyday spaces, this book (re)places the traditional therapeutic landscape within a contemporary theoretical setting (DeVerteuil and Andrews, 2007; Williams, 2007). In seeking to frame these current theoretical concerns against therapeutic experience and performance within a mix of historic and contemporary water-based settings, some valuable connections become apparent (Conradson, 2005b; Wakefield and McMullan, 2005; Lea, 2008; Dunkley, 2009). Within the different watering-places, material forms, curative metaphors and human inhabitations show a surprising resilience and commonality. Individual and communal expressions of health in therapeutic landscapes were shaped by a range of internal (embodied and spatial) and external (cultural, spatial and economic) performances. The discussion of these multiple performances of health lies at the heart of the work. In the mixed settings of the Irish watering-place, the different components of health performance were shaped by and within bodies, cultures, spaces and economies. In addition, in choosing these themes, a prime consideration was their potential wider applicability to other settings around the world.

Approaches

A hybrid approach was used to develop a sound evidence base against which performances of health at the watering-place could be better understood. In the primary and secondary accounts of those places, drawn from a rich mix of historical and contemporary materials, individual health narratives revealed deeper cultural discourses and metaphors. There is insufficient space for a full listing, but the acknowledgments and final bibliography give a sense of their breadth of coverage and format. In visits to historic and contemporary settings, vocal, visual and performed inhabitations brought an additional dimension to the complex ways in which water functioned as a source and medium for health, echoing a notional 'archaeology of discourse' (Gesler, 1993). In using a broad ethnographic approach, primary data collection was carried out via interviews, participant observations and informal conversations. By participating in a number of holy well pattern days, as well as immersions in the waters of sea and spa, I also engaged in direct observer participation in the spiritual, secular and healing performances in place. Finally, in the conversations at these places a more spontaneous yet rich vein of story, narrative, metaphor and *dinnseanchas* (place lore) added an embodied, performative and lived dimension to the visits and a pragmatic richness, to the subsequent critical and theoretical discussions. While the secondary material was drawn from a range of academic journals and books, the *dinnseanchas* of more popular sources, images, oral testimonies and stories were also utilized.

Some important assumptions and omissions should be briefly noted. In placing the therapeutic landscapes of water within wider narratives of healing and well-being, the contestation of their verifiable medical curative power has a long and vibrant history. For many, the spa, well or bath were the playgrounds of the idle, the gullible, the superstitious (Porter, 1999). In every spa, there were 'asps'; slippery quacks and dubious purveyors of water in a desecrated snake-oil form. Those waters are also consumed in place by 'saps', the classic consumer herd for whom dubious and unverified cures were part of 'the hypochondriacal interplay between organic medicine and the half-acknowledged underground realm of the psychosomatic placebo cure' (Porter, 1990, xii). In all of the place-studies, these tensions and impatiences are reflected in how those sites were viewed from scientific medical perspectives. It is important from the beginning, to acknowledge those perspectives within the text. Yet it is on the complex performances of health and hard-to-authenticate ownerships of health in place that this work concentrates. In addition, the utilization of water as a healthy material is deeply entwined in cultural narratives and enactments of folk-medicine and CAM, health paradigms central to understanding the enduring power of such therapeutic settings.

Given the focus of the book, there are a number of additional aspects which could have been considered in relation to water, health and place. All are mentioned in passing though not discussed in any meaningful depth. The first relates to environmental health and the development of clean water (Bergholdt, 2008). While clearly of particular relevance to health, these aspects were omitted due to their pervasiveness and lack of clear connection to all of the theoretical themes developed in the book. A second linked theme was water's link to cleanliness and bodily hygiene and the ways in which this had become embedded in cultural values (McClintock, 1995). Again this is noted in passing, especially in relation to Turkish baths, but is not central to the book's themes. Finally there were connections between water and physical-fitness; recreational and leisure aspects that would certainly fit within a wider definition of health and well-being (Kearns and Gesler, 1998). While space does not permit the fuller development of these three aspects, all representative of a more public health focused narrative of water and place, they represent a potential extension of the research in the future.

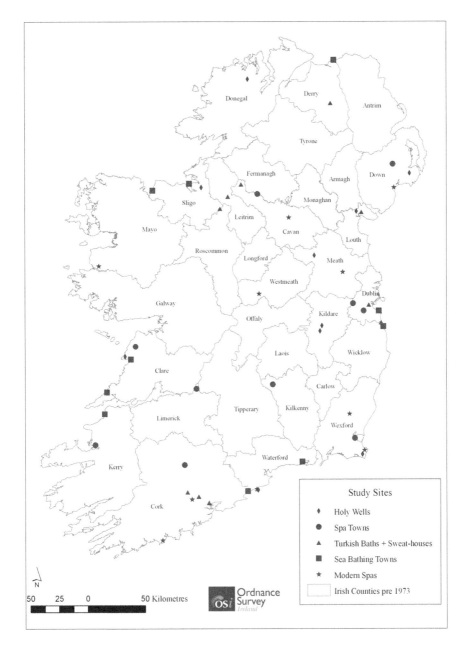

Figure 1.1 Map of Ireland with Detailed Place-Study Locations
Source: Boundary Maps, Ordnance Survey Ireland. © Ordnance Survey Ireland/Government of Ireland Copyright Permit No. MP 008009.

Locations

The first place-study looks at *holy wells*. These types of wells are not unique to Ireland, but find their deepest expression there, where the use and meaning of the healing well sustains into the present-day (Healy, 2001). There are 3,047 officially recorded holy wells across the island, while many more have disappeared over the centuries. Holy wells provide an important psycho-cultural, though contested, link to Ireland's pagan and Christian pasts and act simultaneously as sites of physical, mental and spiritual healing. As therapeutic settings, they have a solidly curative link to place, through the performances of individuals and communities who use them for such purposes. Though much less widely used than in the past, they remain a contemporary presence and a setting not previously studied by health geographers. They also represent natural and symbolic therapeutic landscapes as envisioned by Gesler (1992) and, through their discursive curative dimensions, provide a useful setting for contested health meanings around old/new age spiritualities and wider dimensions of health and well-being.

The second place-study focuses on the *historic spa town*. Ireland, as part of the British Empire, reflected and replicated the curative, social and spatial performances found elsewhere and the peak years of the spa town in Ireland were from around 1680 to the late 1800s (Kelly, 2009). There is one currently functioning spa town, Lisdoonvarna, albeit a relict example. In all there were about 200 identified 'spas', but only around 20 of these developed any sort of commercial function. As classic traditional therapeutic landscapes, they exemplify built and social dimensions of curative place, while also being representative of a nominal 'reproduction of place' notion linked to early forms of globalization through colonisation (Gesler, 2003; Hoyez, 2007a). In providing a new set of grounded studies they also add to a relatively slim existing literature on this watering-place form.

The third place-study considers two unusual linked forms, the *Turkish bath*, and its vernacular variant, the *sweat-house*. Around 50 Turkish bath establishments were developed across Ireland in the mid-nineteenth century, on the back of a renewed interest in hydrotherapy, and some lasted until the 1940s. Though structurally different, a vernacular rural equivalent, the sweat-house, was a primarily a seventeenth to nineteenth century phenomenon, especially in the north of the country. Approximately 279 documented examples remain of this form of healing place (Department of Environment, Heritage and Local Government (DoEHLG), 2008; Northern Ireland Environment Agency (NIEA), 2009). While the role of water in these places is expressed in specifically embodied performances, the study of the bath, alongside the well and spring, is a useful comparative form, linked closely to cultural dimensions of place and health. In the built forms of the baths, there are links to more commercialized expressions of hydrotherapy as a business, which is set against the sweat-house as a natural, pragmatic, communal and vernacular form. These two 'sweating-places' also provide a useful historical link between performances of health at inland and coastal spas (Shields, 1991).

A fourth place-study examines the development of *sea-bathing*, again reflecting a form originating in England, but one which was reproduced along Ireland's extensive coastline. Around 60 such towns and villages of varying sizes were spread around the coast. Some replicated the forms and functions of the larger British and Continental towns, but they also represented a valuable variation in the meaning and cultural construction of the watering place as part of Empire. Some were in place as early as the 1730s while many still have a similar function today. Though their explicit health identities died away in the 1920s, contemporary associations with well-being remain strong. As the natural successor to the spa, such sites have again rarely been discussed in any depth by health geographers, yet have contemporary echoes in discussions on contested models of health at the beach (Walton, 2000; Collins and Kearns, 2007). Within Ireland, and elsewhere, they are also representative of wider social-cultural reproductions of place, while simultaneously marking the second phase and ultimate decline of the traditional watering-place (Kelly, 2009). Finally, the presence of the relatively unique vernacular form of the seaweed-bath provided a valuable contrast to the colonial template.

The final place-study, based around the *modern spa,* represents a contemporary expression of a traditional healing form, albeit aligned to recent globalized developments and leisure and wellness tourism (Smith and Puczko, 2009). Within Ireland, there are over 50 commercial spas of different types and these act as important places in which to study the continuity of meaning, culture and society as expressed through its healing places. In examining the ways in which the spa has assumed new identities, forms and expressions, it also provides a valuable contemporary setting which allows one to identify where narratives and performances of health have departed, returned to and/or evolved from a historical script. In the more contemporary theoretical works on spas and wellness, the setting also provides a previously under-explored setting, at least in Ireland, against which to explore more experiential aspects of the therapeutic place (Lea, 2008; Smith and Puczko, 2009).

The individual place-study chapters are not complete survey descriptions of the specific watering-places concerned but rather a, 'tactical use of comparisons (case-studies) while creating a composite picture' (Mackaman, 1998). Three sites (from a pre-selection of 10) are chosen from each place-study for more detailed comment on summary health-histories and to provide empirical detail. The historical dimension is an important constituent, emphasizing the mobile identities and connective/collective health histories of the watering-place across time and space (Geores, 1998, Andrews and Kearns, 2005). In addition, Strang's elegant description of different water-based settings as being 'typical and unique' should be noted (Strang, 2004). While diversity and difference matter, some commonalities of understanding are also necessary.

The concluding chapter looks across the different place-studies to suggest that while performances of health reflected then current health paradigms, beliefs and cultures, there were strong similarities in their expression. Those identities were shaped by practices which embedded the lasting reputations for healing associated with therapeutic landscapes. Cures and treatments linked to a range of chronic and mental health conditions for example, provide a clear thread across the sites. There are also a number of cross-cutting themes, all set within a performative framework, which recurred in place, such as the roles of gender, colonialism, liminality and the life-course. In addition, there was a strong sense of a moral geography in place across all of the sites, suggesting the need to consider links between spiritual, pragmatic and commercial dimensions in better understanding performance in a health context. The examination of individual performances of health in place recognizes that self-place experiences of health are shaped by wider structural elements, wherein lay and professional beliefs and behaviours are inextricably linked. These experiential aspects are visible in the person, in embodied/phenomenological terms, and in the place, through mobile performances and ownerships. In the first of these places, the holy well, the moral mix of spiritual, pragmatic and contingent cure provides a good starting point, in particular given it is a site that covers both historical and contemporary performances of health.

Chapter 2
Holy Wells: The Faith Cure

Introduction: Wells and Wellness

> The offerings left at wells seem to reflect every need of human existence: for health, for fertility, for jobs, for houses, for a good partner in life, for family harmony. The coming to the well, especially if that entails some hardship, the circling of the 'stations', the adding of stones to cairns, the prayers repeated like a mantra, all of these serve to focus the mind to a degree that allows our own healing powers to come into play, or at least to bring peace to mind and soul ... despite the lack of any official help or protection, they have always belonged to the people, the ordinary folk of the countryside who have no office, power or influence. The cult of the wells could not have endured so long unless it satisfied some deep felt need in our consciousnesses. (Healy, 2001, 116-117)

One of the core traits of a geographer is an irresistible urge to 'read the landscape'. In coming across a setting like St. Kieran's Holy Well (see Figure 2.4), one's eye is drawn to the attractive setting, the rocks and trees and the large stone oratory on the top of the hill. Yet on closer inspection, you find other features, such as rags, red ribbons, notes requesting help and a large metal spoon, the meaning or provenance of which you want to know more about (Bourke, 2001). Often there is a passing visitor, usually local, who can tell you a story (if you ask) about these objects or the various 'cures' associated with the place. The site has not always looked the same (see Figure 2.3) yet the same stories and objects were to be found in the past as in the present place, suggesting the stubborn co-location of therapeutic performance, reputation and inhabitation at this and every other well. To better understand these therapeutic settings, we must move past the gaze and into the heart and body of the space.

Holy wells, as material forms with curative and spiritual reputations, are found across many cultures from Europe to Asia and the Americas (Rattue, 1995; Bourke, 2001; Wilson, 2003). In Northern Europe, they are associated with Celtic and Nordic cultures, with common cultural roots shared across those geographies (Carroll, 1999). As noted in places like Bath and Paris, holy wells have attributed pagan origins with subsequent evolutions as Christian sites (Bord and Bord, 1975; Gesler, 1998). While the pagan foundations hypothesis has been contested, there is a strong association with the less contested pagan worship of nature (Rattue, 1995). Across pagan cultures, there was evidence of the worship of natural elements of the landscape, usually in the form of stones, trees and mountains as well as lakes, rivers and springs (Bord and Bord, 1975; Strang, 2004). These elements of the landscape had, through their ascribed spiritual powers, an associated healing capacity. In particular, the liminal position of the spring or well as marker of the emergence of life from the earth and as a boundary and portal between the chthonic and human worlds, gave springs and wells particular metaphorical curative meanings (Geores, 1998; Strang, 2004).

Holy wells are particularly associated with the island of Ireland and sustain a healing reputation into the twenty-first century; a testimony to their deep cultural histories as therapeutic places. Despite attrition through destruction, loss and infilling, they retain a significant presence in the landscape (Rickard and O'Callaghan, 2001; Healy, 2001). Though more common in rural areas, they are also found in urban settings. In Irish the wells are referred to as *Tobair Naofa* or *Tobair Beannaithe*, translatable as holy or blessed wells respectively. As assumed former sites of pagan worship, the new Christian church in both Britain and Ireland took a pragmatic approach of absorption rather than suppression (Carroll, 1999). This included the tactic of appropriation of well identities from the eighth century on, transferring them from previously pagan associations to new Christian saints, though some of those earlier connections lingered (Rattue, 1995). Within Ireland, they remained significant spiritual sites of pilgrimage and healing up to the mid-nineteenth century (Belhassen, 2008). With the development of a more formal network of churches and parishes from that period, the utilization of wells, and their associated cultural meanings declined significantly, though they never completely disappeared.

While other therapeutic landscapes research has focused on spiritual sites, such as pilgrimage centres, the holy well has not been explicitly considered within that literature (Gesler, 1993; Wilson, 2003). Yet, the rich potential is visible in the extent to which other subjects, especially anthropology, folklore and history, have extensively covered the holy well (Turner and Turner, 1978; Logan, 1981; Healy, 2001; Ó'Cadhla, 2002). In addition many of these texts discuss the health and healing dimensions of the wells and list the cures to be experienced at them (Logan, 1972; Buckley, 1980). They represent fertile ground for health geographers who instinctively focus on material, metaphorical and inhabited connections between health and place. In addition the definitions of therapeutic landscapes as places with natural/built, social and symbolic associations with health and wellness are all fully applicable to holy wells and offer opportunities to ground those connected conceptual terms within living environments (Williams, 1999c; Gesler and Kearns, 2002).

Ritual performances of health in place were expressed through a series of embodied practices at the wells. As well as the ingestive act of drinking the actual healing waters, these included 'rounding rituals', a series of circular routes followed by the visitor in and around the wells that involved prayer and other penitential practices (Carroll, 1999; Ó'Cadhla, 2002). These communally-owned rituals were sometimes referred to as 'stations' (in Irish *turas*), and were usually carried out prior to drinking the water. They could be practiced at any time, but were most effective on specific 'pattern' days (Hardy, 1836; MacNeill, 1982). A modified form of the name for the *pátrún* (patron saint) of the well, the day usually coincided with the relevant saint's feast day and functioned as a local health pilgrimage. Where a well did not have a specific saintly association, other symbolic days, displaying a deep-rooted temporality or seasonality, often with pagan provenance, sufficed instead (MacNeill. 1982; Condren, 2002). These symbolic enactments and ownerships continue to be reproduced in small, private and personal visits and the continued, and crucially, free circulation of holy water from the wells (Gell, 1998; Strang, 2004).

These cultural meanings of the holy well are sustained at both local and national level through a tradition of *dinnseanchas* or place stories, often unwritten and passed down from generation to generation, as a reminder of why narrative and reputation remain central to the therapeutic landscape (National Folklore Collection (NFC), 1934b; Gesler, 1992). The rituals and forms by which the holy wells were given meaning reflected a range of European traditions such as wakes, pardons and kermesses; annual fairs and gatherings, often with religious connotations, which also functioned as sites of social and place identity (Turner and Turner, 1978; Walton, 1983). While often seen as sites of superstition and of necessary surveillance by the colonial authorities in Ireland, in part due to their public assembly functions, they were, as sites of spiritual/physical healing, functional parts of the health care systems of their time (Logan, 1981; Porter, 1999).

The material structures, settings and appearances of holy well spaces contain a number of elements, linked to natural, built and symbolic environments (Gesler, 1992). Holy wells range from literal holes in the ground to grand covered enclosures with walls, steps and roofs (Logan, 1981; MacNeill, 1982; Healy, 2001). Other elements, which, though not found at all wells, are characteristic of many include; additional pools and streams, paths and walkways, statues of the patron saint or other religious figures with whom the well is associated, small oratories, grottos, station markers, trees, bushes, seats, stones and altars. More ephemeral, but no less important elements include a range of left objects, referred to as votive offerings, in a staggering variety of materials with multiple meanings ranging from supplication and anticipatory gratitude to loss and bereavement (Taylor, 1995; Carroll, 1999). The physical locations of the wells, often in settings, which are peaceful, secluded and aesthetically pleasing, also reflect the therapeutic dimensions of contemplation, affect, stillness and restoration (Conradson, 2007; Slater, 2007). While the holy well, as a site of health consumption, was less commodified than later forms of watering-place, such as pilgrimage sites, they were mobile settings inhabited by a complex mix of users, inevitably giving rise to some forms of economic activity around souvenirs, sleep and sustenance.

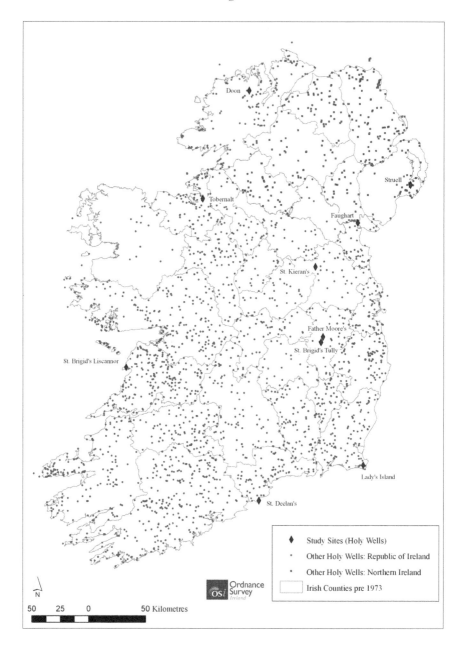

**Figure 2.1 Map of the Distributions of Holy Wells with 10 Place-Study
          Locations**

Source: Boundary Maps, Ordnance Survey Ireland. Sites and Monuments Records (Northern
Ireland: NIEA Republic of Ireland: DoEHLG). © Ordnance Survey Ireland/Government of
Ireland Copyright Permit No. MP 008009.

While a global figure of 3,000 is regularly cited, the total number of extant holy well sites (including bullauns, sacred stones with pools), as recorded in official Sites and Monuments Records for both the Republic and Northern Ireland is 3,047. Even this is acknowledged to be an under-estimate. Geographically, they are spread across all parts of the mainland as well as on offshore islands (see Figure 2.1). From the 10 specific holy wells to be used in the remainder of this section three were chosen to provide 'typical and unique' histories. These were; St. Brigid's Well, Liscannor, Co. Clare (see Box 2.1); St. Kieran's (Ciarán's) Well, Castlekeeran, Co. Meath (see Box 2.2) and St. Declan's Well, Ardmore, Co. Waterford (see Box 2.3). The remaining wells include other Brigidine sites namely St. Brigid's Well and Stream/Shrine at Faughart (Co. Louth) and St. Brigid's Wells (Wayside and Garden) at Tully (Co. Kildare). Wells with links to traditional pilgrimages included Lady's Island (Co. Wexford) and Struell Wells near Saul (Co. Down). Tobernalt Holy Well (Co. Sligo) and Doon Well (Co. Donegal) were wells of considerable standing locally while a more recent example was Father Moore's Well at Rathbride (Co. Kildare). In addition a number of other wells from around the country are referred to where relevant.

Embodied Performances of Health: Healing Bushes and Hollows

Holy wells function as sites of spiritual healing where metaphors of faith and belief are central, both in place identities but also in how healing and wellness are experienced through a range of embodied health performances. At many wells, the symbolic power of that healing is primarily imparted by the patron of the well. As a saint with a national status, pagan origins and discursive reputations as healer and feminist icon, St. Brigid is represented at three wells; Tully (near her monastic home and fire-temple), Faughart (her birthplace) and Liscannor (O'Duinn, 2005). Saint's with more localized catchments are noted at Ardmore and Castlekeeran, through associations with St. Declan and St. Ciaran respectively. Even in those places, which no longer bear a saint's name, there are often imputed associations. This applies particularly to Struell Wells, which were referred to in mid-eighteenth-century accounts as St. Patrick's Well, given their proximity to the saint's tomb (Pococke, 1752). Fr. Moore's Well, is named after a ninteenth-century priest who lived in the district until his death in 1826 (Fitzgerald, 1914). These connections at the well to a particular figure (historical or mythical), are an embodied invocation by person-as-patient to person-as-healer, representing an affective phenomenological connection across time and space (Gesler, 1998).

Box 2.1 St. Brigid's Well, Liscannor, County Clare

With direct links to the pre-Christian celebration of harvest, *Lughnasa*, St. Brigid's Well (*Dabhach Bríde*) sits on an elevated site overlooking the broad sweep of Liscannor Bay in West Clare (MacNeill, 1982). There are traveller's accounts of the well and pattern going back to the seventeenth century (Lynd, 1998). In keeping with the fluid and mobile nature of the holy well, it was originally located on the south side of the road, but was moved and enlarged by a local M.P. and landlord, Cornelius O'Brien in gratitude for a cure in the 1840s. The site itself comprises a well-maintained circular enclosure with paved stones and Liscannor flag seats, containing a large statue of St. Brigid at its centre, raised on a flower bed. Behind this, in the upper part of the sanctuary lie a graveyard and mausoleum as well as a stone cross in some trees to the north east. The main well itself sits in the lower sanctuary, at the end of a spectacularly decorated cave/grotto, about 7 metres long, filled with a large variety of votive/left offerings.

Figure 2.2 Left Objects, St. Brigid's Holy Well, Liscannor, County Clare, 2009

Performances at the site are represented by a series of points and paths in the microlandscape, which act as markers for the stations or rounds, the composite penitential/curative rituals carried out to enable the cure to work. Traditionally, one undertook parts of the round on one's knees, while five sets of prayers were said at the statue, on the path above and at the cross respectively, before the well water was drunk three times and the favour requested. Specific cures included those for rheumatic pains, sore eyes, ulcers, violent headaches or 'teeth-aches' (National Folklore Collection (NFC), 1934a). As well as the reputed cures, the site currently has a new healing emphasis around Memorial/Reconciliation/Grieving contained in notes and letters. The pattern at Liscannor, which had a regional reputation, was traditionally performed on Garland Sunday (last Sunday in July) but with other celebrations on 1 February and 15 August. The pattern usually took place on the night before Garland Sunday, with the rounds carried out at midnight and the rest of the night given over to conversation, music and dancing.

In particular Aran islanders from Inisheer would come over on *currachs* and sing through the night. Once the night was over all would head down to the beach at Lahinch for horse-racing on the strand. The site was considered very well-behaved compared to most. Logan noted that the well had been thoroughly Christianized by the late 1970s (Logan, 1981). Local/generic splits in spiritual identity can be observed, with a partial make-over in the First Marian Year of 1954 when statues of Mary were added to wells in the act displacing local saints such as Brigid (Taylor, 1995). Finally, the position, and increasing cultural consumption of the well by passing tourists en route to the Cliffs of Moher, one of the country's emblematic tourist sights is noted, potentially problematically, in relation to the recent inclusion of the well into booklets sold at the latter site.

While Lady's Island, a fifteenth-century pilgrimage site, had a more generic healing role, more specific embodied cures were listed at all of the other sites (Logan, 1972; Bord and Bord, 1975). Such cures typically took two forms, as either a cure attributed to the whole well, or linked to a section of the well. So for example eye cures, a common theme across the island, were recorded at Ardmore, Liscannor and Tobernalt, while at Struell, there was a specifically named Eye Well at the centre of the site. In focusing on a particular body part, that embodied identity was reinforced, while more spiritual dimensions of such a cure suggested a repaired/ improved ability to look out from within (Varner, 2002). Several wells were noted as having specific cures for 'mental (health) problems'. At Tobernalt, the setting against a steep stone cliff and within a grove of trees close to the shores of Lough Gill, give it a strongly natural therapeutic flavour. There was a leper and insane colony at a nearby island in the seventeenth century, which was granted access to the well site at restricted times for just such therapeutic benefits (Boylan, 2002). Cures for back problems were recorded at Ardmore and Tobernalt and Castlekeeran. At the latter, there was a natural hollow in the limestone, fitted to the curvature of the back, where one could literally, bodily inhabit the cure. Cures for rheumatism and general mobility problems were recorded in four of the wells as were headaches and toothaches. Other cures recorded at the representative wells included those for warts, sprains, hearing and paralysis. Castlekeeran had an additional reputation for veterinary health. Horses were brought annually to bathe in the stream that ran in front of the well to provide them with physical healing and spiritual protection and reflected records of animal cures at wells across the British Isles (Bord and Bord, 1975). Finally, relationships between cure-place were visible in whooping cough cures identified almost exclusively at city wells, a curative link to be expected in a primarily urban disease (Logan, 1981). Perversely, the presence of so many cures for rheumatism and eyes in rural areas suggest that rainy exteriors and smoky interiors created a reflexive relationship between environments and health need (Logan, 1972).

Mindful of Conradson's notion of the therapeutic landscapes experience, that experience was expressed in embodied performances (Conradson, 2005b).

A common enactment was the attachment of rags to a nearby tree, whereby the patient in turn; dipped a rag in the well, rubbed it on an affected body part, and then tied the rag, in a symbolic expulsion of embodied illness, to a nearby tree or bush (Simon, 2000). As the material object faded in time, so too did the physiological complaint. Almost identical practices were noted in nineteenth-century Sudan (Wood-Martin, 1892). This specific performance of health was one of a number of embodied practises around water at the well, wherein; 'meanings encoded in it are not imposed from a distance, but emerge from an intimate interaction involving ingestion and expulsion, contact and immersion' (Strang, 2004). While drinking and dipping in the water were the most common forms, at Struell large bathing pools for men and women represented a parallel immersive ritual. There were also significant affective dimensions in those embodied experiences of health in the shock of the cold water and the emotional transfer of illness from subject to object. Indeed, in a particular locational act of 'shock therapy', patients at Kilbarry in Co. Roscommon were locked up in a covered space for three days. In an echo of the cure at Epidauros, they connected in this dark, oppressive space, with their deeper 'other' self through dream, emotion and doubtless, deep trauma as well (Logan, 1981; Gesler, 1993).

Embodied performances of health in place had additional symbolic meanings. The rounds, whether carried out the easy (on shod foot), or hard way (on bare foot or knee), represented a specific directionality of movement but also a physical and punitive contact with the earth. At Liscannor, part of the rounding ritual involved walking around a large cross above the well, keeping your hand on the cross as you walked. At Ardmore, Croker recorded in 1818 the consumption of earth from St. Declan's Grave, visited then as part of the pattern ritual; a rare example of geophagy as cure in Ireland. These macabre tendencies were also found at Faughart Well, where the water was traditionally collected and symbolically drunk from a human skull. In addition, the water was taken away and used symbolically as a prophylactic in houses, where it assumed a preventative health function. As noted previously, rag trees/bushes (or 'cloutie trees' in Scotland) had additional embodied meanings (Bord and Bord, 1975; Healy, 2001). Traditionally rags were red, a common colour for shawls and petticoats along the west coast of Ireland. Over time any material was used to provide a metaphoric transfer of illness from person to tree, but a key aspect was the need for that object to be worn and by extension have a physical contact with the (assumed) diseased body. Thus a symbolic act of detachment and in a sense, disembodiment took place, while objects such as coins and clips and pins, which held clothes in place, were also popular items for this act of metaphorical separation (Carroll, 1999). At Tobernalt a variant on the performance saw visitors taking away existing rags as 'sacred charms' and 'leaving others in their stead to suck new virtue from the place' (Henry, 1739, 11).

At Father Moore's well the cure and connection with mental health were embodied by proxy. After his death in 1826, one of the relics left behind by the priest was his silk hat, which became embodied with his reputational healing power (Fitzgerald, 1914). The same hat is in pretty frail condition now, but is still brought out every August where hundreds of visitors queue to wear it. This may be easily dismissed as a simple and superstitious story of gullible country people, but the reputational aspect of healing object in place is telling in both its narrative and ongoing resonance in 2009. It is also a direct example of an embodied performance whereby the modern patient can quite literally 'step into/try on' the clothes of the holy healer and in that way physically absorb the power of the cure (Logan, 1981). Other examples of this embodied connection to the spiritual provider of the cure were the regular naming of hollows in rocks or stones as the saint's 'finger prints' or knee prints' such as at Faughart Shrine, St. Kieran's and Tobernalt. At the latter well, visitors can connect to the spiritual in a directly embodied form by placing their own fingers in the grooves (Logan, 1981).

More widely the spiritual healing power of the well was sometimes in conflict, but generally co-existed, with vernacular narratives of health and wellness in place. The physical connection between the curative elements of the wells and their surroundings was discursively created and owned. The curved rock at Castlekeeran and Father Moore's hat did not of themselves have a medicinal power that one might see in an object like a pill or ointment, though these too can be seen as discursive creations (Heller, 2005). Rather, it was the texts and discourses, either through the sign at Castlekeeran pointing out the 'Chair for Ailments of the Back', or more commonly, the repetition of 'success' stories around Fr. Moore's Hat, that provided the meaning to the cure. Even I have engaged in that process. At Fr. Moore's Well, the Sunday after the hat was produced in August 2008, I met an elderly man, walking around the well with his two grandchildren who assured me he had been cured in the intervening week of a chronic back problem after wearing the hat and drinking the water at the well. He seemed pretty sprightly to me. The point here is not the scientific evidence of the cure, but rather the belief in the power of the hat to heal and the passing on of that story to another who in turn repeats it. Here the embodied, narrative, symbolic and spiritual elements of a cure experienced at a therapeutic landscape all combine in the accretion and reproduction of its reputation. As Taylor attests while talking about holy wells in general, (with an added health slant in italics):

> The stories preserve the enchanted (*symbolic therapeutic*) landscape; the well and other places like it literally hold the stories. Those raised with the stories need only hear the name of a given place or else see it to rehear the story that is tied to that place – or at least know that there is a story that someone else 'has'. Perhaps it is less a question of whether or not one 'believes' the story than of a (*therapeutic*) landscape that has (*health*) stories in it – stories it can tell you – versus a landscape that talks about other things or is mute. (Taylor, 1995, 61)

In the spiritual dimensions of the therapeutic ensemble, where one's own recurring presence in a healing place draws from others who have been there before you, the embodied traces of a curative past are preserved in the stories and cures in the place, linked to reproduced narratives and re-performed lived actions (Lorimer, 2005).

These broadly phenomenological associations also reflect symbolic and natural components of therapeutic landscapes (Gesler, 2003). As places of healing, the narrative connections of health to place were solidly spiritual and all the wells studied were examples of 'sacred landscapes' (Taylor, 1995; Conlon, 1999). In noting that 'healing and holy have an etymological kinship', McKinlay suggested that, 'we need not wonder that healing wells were, as a rule, reckoned holy wells, and vice versa' (McKinlay, 1893). Despite occasional attempts to test the waters for their mineral content, it was the symbolic religious/spiritual healing power that was the core component of this particular type of water cure (Bord and Bord, 1975; Harbison, 1991). While comments on the 'ordinariness of the water' at Ardmore were meant to debunk any scientific reputation it might have, health meanings were sustained through lived and experiential engagements between the human and the place (Hardy, 1836). The meanings drawn from that engagement were of a direct perceived experience of healing, even if that was only a placebo effect (Gesler, 1998). In the case of visitors I encountered at other wells, such as Doon and Faughart (where a visitor had come from near Newry to treat a persistent eye problem), their stories of cures, or their own presences in search of a cure, were direct examples of a belief in the power of place as material site of health and healing. In their engagements with water, with and through their own bodies, the therapeutic dimensions of place were, for pilgrims and other users, experiential and just as importantly, enacted. While these self-landscape interactions were enacted in place, they were additionally shaped by communal and vernacular narratives to create both a physical and imaginative curative experience.

Cultural Performances of Health: Sacred and Profane Rituals

Health and 'well'-ness were reflected in, 'the simultaneous and ongoing shaping of self, body and landscape via practice and performance' (Wylie, 2007, 166). Yet this was mirrored in wider cultural performances of health central to understandings of the therapeutic landscape of the holy well (Gesler, 1993; Williams, 1998). As sites, they are open, and anyone, whether catholic or not, are free to visit a holy well. Such visits continue to take place on a daily basis. While some commentators express concerns that many visitors do not understand the 'true' meaning of the well, this is in essence a concern with 'authenticity'; a theme often encountered in tourism and leisure geographies (Urry, 2002; Timothy and Olsen, 2006b). While appreciating the spiritual and healing dimensions of the well, it should also be accepted that individuals have their own understandings of the well; even if that is an interest solely in the 'kitsch' factor of the left objects at Liscannor or the rag trees at Tobernalt or Faughart. In addition, part of the rationale of this section, is to discuss precisely those 'inauthentic', even 'unhealthy', performances at the well. Some of these practices were akin to the carnivalesque, and pointed to a certain contestation, beyond existing biomedical critiques, of the healing status of the well (Bakhtin, 1984). These culturally constructed and more importantly, collectively performed activities were most fully expressed in the often profane activities of the pattern day.

While a number of debates have emerged about the precise form, timing and causality of the activities of the pattern, variants on this theme are certainly recorded around medieval pilgrimage sites as well as at holy wells in Britain (Bord and Bord, 1975; Turner and Turner, 1978; Bremer, 2006). Mirroring similar non-Christian performances in India and Morocco, Carroll and others argue that 'rounding rituals' at holy wells date from the Counter-Reformation period after 1605, peaking in the eighteenth century (Rattue, 1995; Carroll, 1999; Sellner, 2004). At Struell, Liscannor and Tobernalt, visitors assembled the evening before the relevant day, usually celebrated the spiritual rituals by performing the rounds at midnight, and then partied on into and usually through the night (Logan, 1981; Rickard and O'Callaghan, 2001). Variants included an early morning start on the pattern day at Ardmore. The form of the rituals was prescribed at each well and there was considerable variety in the spiritual acts, which had to be performed. The rounds were often carried out on one's knees, a relatively painful process on rough ground. The rounds were repeated at different intervals depending on the well; 15 times at Liscannor, five at Castlekeeran, three at Ardmore and Tully (National Folklore Commission (NFC, 1934a and 1934c). Prayers such as Pater Nosters, Ave Marias and decades of the rosary were said in variable amounts at each of the stops on the round. Traditionally, the water was accessed at the end of these regulated enactments, where it was either drunk, carried away or used on the body via a dipped cloth or rag. Having performed one's spiritual and associated curative health duties, one could then relax and concentrate on the more social aspects of the event. In recently revived pattern rituals, there are similarities in terms of prayers, timings and the consumption, collection and blessing through water. Even though many of the penitential and indeed, social excesses of the past are avoided, the ritual cultural metaphors and performances remain the same.

Historically, the rest of the evening was then given over to the more socialized, and for many, attractive, rituals of talking, eating, drinking, singing and dancing through the night. At Liscannor (see Box 2.1), the popularity of the pattern was partially due to the anticipated attendance of fine singers from the nearby Aran Islands as a special attraction (MacNeill, 1982). Elsewhere, the presence of stalls and tents at Castlekeeran (see Box 2.2) and Ardmore (see Box 2.3) were noted, all selling a variety of souvenirs but also food, rough whiskey and even rougher *poitín,* a dangerous (and now illegal) concoction distilled from bran and potatoes. At Ardmore, Hardy, a distinctly hostile witness from 1836, noted that:

> ... the sanctity of the day ... did not prevent its night being passed in riot and debauchery. The tents which, throughout the day, the duties owing to the patron saint, caused to be empty, at evenings became thronged with the devotionalists of the morning, and resounded till day-break with the oath of the blasphemer, and the shouts of the drunkard. (Hardy, 1836; cited in Ó'Cadhla, 2002, 26)

At Struell, there was a particular concern with the penitents' behaviour, who plunged naked into baths undifferentiated by gender and assembled on the top of a nearby hill and made merry, taking advantage of local narratives that, for that one night only, no further sin could be contracted (Harris and Smith, 1744). The degeneration of patterns into violence and 'faction-fighting' was also a common feature (Ó'Cadhla, 2002). There are a number of recorded examples, especially at Ardmore and Tobernalt, while at Tully; 'There was a famous pattern held at this spot over 150 years ago, but the patterns fell into disrepute for want of piety.' (NFC, 1934c, 40). This 'want of piety' and in particular the combination, as described in Croker's 1810 account of the Ardmore pattern, of 'bloody knees from devotion, and bloody heads from fighting', represents succinctly the juxtaposition of the sacred and profane (Ó'Cadhla, 2002). The factions tended to be groups of men from nearby villages, rural districts or urban neighbourhoods, for whom the pattern was a rare opportunity for public assembly, which, combined with the presence of groups of women, led to an explosion of alcohol and testosterone-fuelled behaviours (Evans, 1957). In addition, it seems as if faction fights were ritual performances, as in Ardmore, where they were re-enacted between the same gangs annually across a 40 year period (Ó'Cadhla, 2002). Nonetheless, at Ballybunion in County Kerry, a faction-fight on the beach after a pattern in June 1824 led to the deaths of 20 men so the violence was genuine. Ó'Cadhla notes this sacred/ profane mix and its 'carnivalesque' dimensions, to describe similarities between the performances around the Ardmore pattern and the global pre-Lenten carnivals of Europe and South America (Ó'Cadhla, 2002). This also tallies with accounts of connections between holy well patterns and older pre-Christian festivals especially at Liscannor (Evans, 1957; MacNeill, 1982). Some of these performative aspects of the carnivalesque, in particular its allowance of the liminal and the suspension and subversion of normal time and roles, are useful notions when applied to other healing sites later in the book (Bakhtin, 1984; Turner, 1977).

What the juxtapositions of the sacred and profane did suggest however, were the close association of curative and cultural performances of health in place, even if those were sometimes contradictory, especially in relation to the moral geographies of behaviour at the well (Eliade, 1961; Healy, 2001). While Carroll usefully identified distinctions between curative and penitential performances at wells, the communal remained a significant element (Carroll, 1999). This echoed the anthropological term 'communitas', where commonalities of meaning, intent and behaviour were expressed by groups in place in many healing landscapes across the world (Turner and Turner, 1978). When people were ill, or concerned about a health outcome, either good or bad, it was important to share that burden or joy with others. As a psycho-cultural aspect, the notion of 'sugaring the pill' was often acted out in place, as incentive, reward, support or release of tension. Thus the spiritual healing aspects of the pattern, containing elements of pain and endurance, needed in a sense to be balanced by something more pleasurable; all the more so when in company. Similar contradictory performances of spirituality/penitence and socialization can be observed in contemporary settings like Pamplona or

Lourdes (Nolan and Nolan, 1992; Taylor, 1995). Whatever the motivation, many therapeutic places contained this mix of medicine and pleasure. The pleasure aspects, outside of the all-night music and dancing, took a variety of forms. The tradition at Liscannor was to repair to nearby Lahinch beach the morning of the pattern for organized horse races while many Sligo residents traditionally took a boat out from the town to Tobernalt on pattern day and afterwards picnicked on small islands in Lough Gill. Around the patterns at Castlekeeran (recovering) and Ardmore (very much alive) there was a week of festival entertainments and activities. At Tully, celebrations incorporating craft workshops, fire rituals and creative dance have been part of a week-long event organized by the *Cháirde Bhríde* (Friends of Brigid) collective since the late 1990s, while a similar collective organize events at Faughart.

Box 2.2 St. Kieran's Well, Castlekeeran, County Meath

Set within a micro-landscape of tree, stone and water, St. Kieran's Well (Figure 2.3) in Castlekeeran, County Meath, was described as 'one of the most beautiful holy wells in Ireland' (Wilde, 2003, 141). It lies close to an eighth century monastery associated with St. Kieran and had therapeutic spiritual associations as early as the 1200s. (O'Connell, 1957). The first formal reference to the well's use was in the mid-seventeenth century (Rutty, 1757). The site itself is about 80 metres square and includes a range of classic holy well elements. It runs from a steep hill at the back, containing a corbelled oratory, down to a flatter front section containing a rag tree, a double well, a small rivulet, a main healing well and two entrance bridges crossing a fast flowing tributary stream named after the saint (NFC 1934c). The wells are natural pools within exposed blocks of limestone. In 1839, there was a rumour that a tree overhanging the well was flowing with blood, which led to large numbers flocking to the well in a practice more reminiscent of Mexico or Italy. The well is now shaded by a smaller tree and has five silver metal crosses dotted around it as well as a large spoon on a chain for drinking. While the main well was broadly curative, there is in the limestone rocks a place for curing 'ailments of the back'. At the double well, cures for headache and toothache depended on the side of the well the water were taken from. People also washed their feet in a rivulet with reputational cures for rheumatism and corns (NFC, 1934c). The rag tree contained an assortment of votive offerings and included rags, cards, crucifixes and medals. Behavioural taboos at St. Kieran's dictated that it could not be used for any purpose other than drinking. St. Kieran's also had three holy fish, which by reputation only appeared at midnight on the pattern day. If these were taken, they usually returned mysteriously to the well and brought bad luck to the thief (NFC, 1934c).

Figure 2.3 St. Kieran's Well,   Figure 2.4 St. Kieran's Well,
            Castlekeeran, Co. Meath,   Castlekeeran, Co. Meath,
            c1901   2009 Pattern Day

Source: Courtesy of the National Library
of Ireland.

While the period around the saint's birthday from 14 June to the 23 June remains a
popular date for visits, the established date for the performed pattern is the first Sunday
of August. A visit to the well at midnight of the previous night has recently been revived,
a direct echo of earlier celebration. Large crowds then assembled at the site on that
Sunday to meet and pray. There would be a mix of people of different classes and in the
past would have had its mix of tents and all night carousing. There was also a traditional
recreational element with a challenge Gaelic football match played, usually between
Meath and Cavan, which attracted large crowds. Photographs from the turn of the
twentieth century record crowds of 6-7,000 attending the pattern. Finally the well is also
a contemporary site of visitation for neo-pagan groups and white witches. While these
conflicting understandings of formal versus informal sacredness can cause contestation,
through the occasional removal of pagan decoration, the performed inhabitations of the
site remain open and accessible.

While specific cures were attributed to wells, their healing powers were also
reproduced in distinctly cultural terms. The linkage of metaphor and health as
noted in the therapeutic landscapes literature, merits reconsideration here,
especially in reputational connections to the spiritual (Geores, 1998; Gesler and
Kearns, 2002). The signifiers of the sacred were visible in all of the wells, such as
the reproduction of the St. Brigid's Cross (a square with a lozenge shaped centre
and angled protruding arms), a widely understood symbol within Ireland. In that
sense, Brigid's reputation as a healer was produced and reproduced, especially
at Faughart and at Tully Garden, as a semiotic object simultaneously framing
and naming sacred and curative space (Smith, 2003). The more ritualized and
arguably contingent symbolic connections between religion and healing were
also reproduced in such places. At sites such as Liscannor, Tobernalt, Fr. Moore's

and Faughart there were detailed instructions on the religious rituals, specifically the rounds or stations that should be performed. The drinking of the water was primarily enacted in these rituals; signifying a relationship between practice and cure and a moral geography that linked the symbolic and the curative.

In another sense the mind and spirit played a role in healing in place, filtered through an instructed and culturally understood engagement with spirituality in place (Holloway, 2006). To receive the power of health in these places it was necessary to respect and be mindful of the spiritual presence that offered that healing. Disrespectful uses of the water were associated with an anti-health outcome, such that for example, at Liscannor, interference with a sacred eel (sacred fish were common elements) would lead to death within three days, weeks or years (Rickard and O'Callaghan, 1981). Given that belief in the cure lay at the core of the well's healing reputation, it then became an individually owned and negotiated act of faith healing within an associative 'faithscape' (Shackley, 2001; Singh, 2006). Yet, during contemporary visits, at Fr. Moore's, Tobernalt and Tully, conversations were held, cures discussed, body parts rubbed, prayers uttered and a less overtly reverential engagement held with cure in place. These actions represented the more socialized and communitarian aspects of health enacted in place (Turner and Turner, 1978; Ó'Cadhla, 2002).

Over time these older performances have been sustained if slightly altered, within contemporary cultures. Some of the new forms of objects found on trees around wells, such as hair scrunchies, plastic slides, bracelets and babies soothers, as noted at Liscannor, Faughart Well, Tobernalt, Tully and Doon, all meet embodied criteria. Other offerings left at wells include a large number of cards, candles, photographs, statues and rosary beads all of which have affective (and implicitly embodied) personal meanings (Bourke, 2001; Healy, 2001). There are stuffed penguins (Faughart) and teddy bears (Liscannor) and at Lady's Well on Lady's Island a decorative pair of pink furry dice. While the specific meanings of individual offerings are personal, there has been a shift to a newer memorial function. Indeed Liscannor's proximity to the tourist Mecca, the Cliffs of Moher has led to many recognisably Irish-American offerings, including a mass card for an NYFD officer killed at 9/11, visible at the well in 2008. Stones, symbolically and spiritually representative of the earth, and reflecting pagan traditions, are left on altars at Lady's Island, while at Ardmore, there is a tradition (repeated at other wells) of scraping a small white cross into the stone at either the well or ruined church beside it (Graves, 1948). At the holy well, the pattern and the small individual and communal rituals augment the more material consumption of the water (in itself of course, a symbolic act) to cement these cultural performances of healing in place (Brenneman and Brenneman, 1995; Taylor, 1995).

There is a distinct connection in how therapeutic landscape experiences are in turn affected by cultural performance in place. There is a communal performance in bodies that circulate amongst water, tree and stone; objects are touched by subjects; bodies are rubbed and bent; nature is decorated by human hand. While some of these actions represent what anthropologists call gift-exchange, where symbolic acceptance of healing and wellness is reciprocally repaid, they are also more than representational (Taylor, 1995; Lorimer, 2005). These enactments are tempered and shaped by their social/cultural contexts, which in turn adds new enactments, both health and unhealthy. While the broad forms of the rituals carried out at wells still retain a healing meaning, subtle changes over time also mirror cultural shifts in how health itself is understood and performed. This also illuminates diversity and difference in those health/ place performances and;

> in terms of its emphasis on how repeated encounters with places, and complex associations with them, serve to build up memory and affection for those places, thereby rendering the places themselves deepened by time and qualified by memory. (Harvey, 1996, cited in Cloke and Jones, 2001, 651)

They may also mark a shift in public perceptions from simple chronic and afflictive injury and disease to more holistic notions of a mind/body/spirit health. These newer meanings are acted out in expressions of memory, loss and personal need (Holloway, 2006; Einwalter, 2007). New age spirituality, where pagan-nature elements are worshipped, also finds expression at the well. Such shifts in ritual performance reflect a movement away from more constructed sites of vernacular or church-shaped spiritual health.

Finally, the carnivalesque social elements represent an additional layer of performance. As noted by Nash, some spatial performances have the feel of a choreographed dance (Nash, 2000). One can see the healing and social aspects of holy well visits as being more akin to a long running stage play, a cultural text that is directed by different directors at different times, though self-direction is also possible. Indeed, at Tobernalt in 1958, the site was explicitly used for a performance of the life of Bernadette of Lourdes (Boylan, 2002). Like a theatre the holy well setting has props, sets, actors that, while sticking to a broad script, can be interpreted and more importantly performed differently in time and place to produce different (health) outcomes through the affective response of its audience (users). All of these often contradictory cultural performances of health remain however, grounded in spiritual elements of faith and belief.

Spatial Performances of Health: Pilgrims and Pagans

Geography also played a significant role in shaping the identity and utilization of the holy well. In what can be termed spatial performances of health, the locations and settings of the wells helped shape their accessibility and use by a range of people (Joseph and Philips, 1984; Smyth, 2005). Carroll makes a case for use by the 'better' classes in society, recounting the 'vast throngs of rich and poor going to Struell' (Harris and Smith, 1744; Carroll, 1999). In addition, their functions as pilgrimage sites also attracted wealthier Catholics and curious outsiders (Nolan and Nolan, 1992). For much of their history however, holy wells attracted a more humble clientele described pejoratively in colonial traveller's accounts as the activities of a 'superstitious peasantry' (Hardy, 1836). There were early seventeenth -century accounts of large groups of poor urban dwellers visiting St. Patrick's Well in Dublin (Carroll, 1999). The volumes of use were considerable, with a widely used figure for Ardmore of 15,000 (Croker, 1824). Figures of 6-7,000 were recorded at Castlekeeran, while 'great numbers' of probably equivalent size were noted at Struell in the eighteenth and nineteenth centuries. Even as recently as 1978, 20,000 pilgrims visited Lady's Island on 15 August. Well worship declined with the development of a more formal church structure and greater affluence in the later nineteenth century changed tastes, fashions and social aspirations (Carroll, 1999). But the holy well has not died away completely and sustains a small but passionate following. Even on a freezing night on 31 January 2008, a group of around 100 assembled at the Tully Wells, performing a number of rituals and ceremonies at the two sites.

Visitor bases for holy wells were primarily local, though bigger centres had wider catchments. Historically people came from the parish and had local affiliations with the well with a nominal catchment of 20 miles (Harbison, 1991). Doon for example, was a site of particular reverence to the people of Donegal, aided in part by the proximity to Doon Hill, the coronation site for the local O'Donnell clan. Fr. Moore's Well, despite its relatively recent history, had and has a strong 'following' within County Kildare. Other wells drew their visitors from regional and national catchments. Here the relative reputations of the therapeutic landscapes, for spiritual and physical cures, were expressed in spatial terms. Liscannor was much frequented by the Aran Islanders (*Aránachs*) and 'Red Petticoats', rural female visitors from nearby maritime counties (MacNeill, 1982). Ardmore was recorded as having visitors from all over the province of Munster for its all-night vigil, while Struell's fame was reflected in accounts from 1836 of a man who had walked there from Galway (a distance of 250 kilometres) in his bare feet, occasionally taking the weight off by switching to his knees (Hardy, 1836; Bourke, 2001). Lady's Island, given its long-held position as an important pilgrimage site, had and still has a national catchment. A more aspirational and culturally constructed form of catchment was associated with Faughart Shrine, where a 'national' pilgrimage was created by the Catholic Church in 1934. This was materially represented by the building of a new covered Shrine in the upper section of the site and a grand inauguration by then President, Eamon de Valera. However the timing in the mid-1930s was also a symbolic expression of a specific agenda of co-dependant church/state nation-building, where health and the cure had a reduced role (Foster, 1988).

Box 2.3 St. Declan's Well, Ardmore, County Waterford

The village has Christian links going back to the sixth century, and is recorded in the twelfth century as a well-established monastic pilgrimage site dedicated to St. Declan (Ó'Cadhla, 2002). The reputation of the town and the holy well developed in tandem. In the nineteenth century, enormous numbers, 15,000 in 1810 and 1840 are recorded for the week around the pattern, 24 July (Hall and Hall, 1843). Specific curative performances involved a visit to three core sites: firstly to Declan's bed, a stone at the low tide mark, under which people crawled to absolve themselves of their sins; followed by a visit to the ecclesiastical site to ring St Declan's Bell; and finally the visit to the Well (see Figure 2.5). Earlier stops on the circuit included a visit to St. Declan's grave where one ate the earth or took it home, though this part of the pattern not surprisingly, lost favour.

Figure 2.5 St. Declan's Well, Ardmore, c1909
Source: Image from the photographic archive of Irish Heritage Giftware, info@ihpc.ie.

In curative terms, the well itself had a generic reputation, though eye cures were specifically mentioned, as were 'sprains, wounds and rheumatism' (Logan 1981). Specific cures were associated with the different sites in the town. On the beach, the stone was excavated at low tide and people crawled under it for a form of spiritual healing, to confess their sins and to cleanse for the subsequent stages of the visit. In addition it allegedly cured backs, though the ritual performance associated with it suggests a quite opposite effect. At the well, the physical, material and embodied acts of dipping in a cup and drinking the water were an essential part of the healing process, with both ingestion and symbolic immersion (through rubbing), as the forms by which the healing water became embodied (Ó'Cadhla, 2002). At the well people drank the water, and did three rounds in a clockwise direction while saying a variety of Our Fathers, Aves and decades of the Rosary. Once completed the people would descend to the town to eat, drink and in many cases, become more than a little merry. The generally profane activities of the day were recorded in a variety of accounts with faction-fighting mentioned in particular. Music was important as well, as Ardmore was in an Irish-speaking area with strong cultural performative traditions (Broderick, 1998). There was no great class distinction to its clientele though the 'peasantry' were occasionally referred to by more elitist commentators. Many people walked to the well or by horse and even boat from the nearby Ring district. Due to its established pilgrimage, visitors came from the wider province of Munster as well (Croker, 1824). Currently, a small assembly of local people gather at the well at midnight on the 23rd to say prayers and welcome in the day. The 'branding' of the saint's name was also an example of how entrepreneurship was developed both by the church and then in a more secularized religious form. The name of Declan is and was everywhere and the linking of the name with place identity is expressed in a variety of settings within the contemporary town.

Travel to the wells was often difficult, given the relative remoteness of some of the sites. This reflected their spiritual dimensions, as liminal places of retreat and reflection. Despite this they were reached through a variety of means, some as alluded to earlier, more painful than others. As a visit to a well was also a spiritual journey, a certain amount of penitential pain was only to be expected. Given the anticipated healing benefits, they could also be considered as *health pilgrimages*, a term rarely used in the literatures. Walking was always a main way to reach the well, especially for the poor. Those who could afford a horse, or better still, a horse and cart, would use that. As other transport networks developed and roads improved that accessibility was widened further while the role of the bicycle, an under-regarded transport mode, opened up access from the later decades of the nineteenth century (Duffy, 2007; Towner, 1996). Old photographs of wells as at Doon (see Figure 2.6) and Castlekeeran show bicycles parked in the background, as well as some very early models of the motor car. The boat, another under-estimated mode, was used at Tobernalt, Liscannor and Ardmore in a relational connectivity across and towards water. Experiences of accessibility and utilization were different, and much more difficult, for the invalid and more generally infirm. As transportation modes developed, access for these disadvantaged groups improved but the rough terrain, and indeed difficult rituals remained a problem. For many more able-bodied visitors, part of the role of the visit therefore, was to act as a disembodied proxy for their ill friend or family member. This role of the proxy, who absorbs the penitential role on behalf of others and who brings back the healing product, in this case holy water, is also an established function of pilgrimage sites like Lourdes (Turner and Turner, 1978; Gesler, 2003).

Diverse users have, through history, marked variant, even deviant use of the sites, while new users represent changing models of utilization at the wells. As an officially recognized ethnic minority within Ireland, the traditionally nomadic Travelling Community have a particular regard and reverence for the holy well (Haggerty, 2007). This reflects the reverence expressed by other Gypsy communities at water-based sites such as St Marie-de-la-Mer in the Camargue. In a sense Traveller's, like other Gypsy, Roma and Sinti groups in Europe, live outside ownership in a sort of migratory commons and so come into contact, and inevitably, contestation with the settled community. Many of the global discourses of exclusion apply fully to how Traveller's are perceived and treated, and may be reasons for a reverence for the holy well. Excluded from both spiritual (by a generally disengaged church) and medical (through limited access to health care services) settings, Traveller's gravitate to those healing places which were open to them and for which they have had a long cultural attachment. In Kildare, both Tully, and especially Fr. Moore's remain very popular sites for Traveller devotion and visits while Struell has also been mentioned in Traveller accounts (Haggerty, 2007). At Tobernalt, Traveller's unique performances, involving wading through the stream in their bare feet, reflect a particularly intense devotion and connection to a cultural cure and faith. At Our Lady's Island in September 2009, at least a quarter of the 2,000 or so visitors to the last night of the pilgrimage were Traveller's, though there was no evidence of the traditional practice of walking around the island with one foot in and one foot out of the water (De Vál, 2007).

The ascribed pagan antecedents of the wells have also made them sites of reverence and meaning for New-age and neo-Pagan groups. Again small examples of resistance/conflict around authentic ownership and meaning have been recorded at both Castlekeeran (see Box 2.2) and Tully. However, there is a relative openness to such groups, consistent with shifting meanings and more importantly free access to the sites. From a health perspective, these new utilizations also point to a strong relationship between place and new age spiritual/healing communities, most fully expressed at places like Sedona, Arizona, often described as the New Age Lourdes (Timothy and Conover, 2006). Tully is a regular port of call for a number of neo-pagan guided tours to Ireland, especially given the associations with St. Brigid, a strong feminist role-model with pagan associations (O'Duinn, 2005; Condren, 2002). The holy well also function as sites for alternative visions of a more holistic healing, echoing some of the discursive shifts to be found in CAM and their relationships with orthodox biomedical views of health (Williams, 2007; Lea, 2008).

The therapeutic qualities of place are also central to affective experiences of health. All of the wells are in natural landscape settings and are isolated and remote. Doon is two kilometres from the nearest minor road, while Faughart Well lies on a desolate hillside. Struell is set in a secluded valley in the rolling drumlin-belts of East Down, while Liscannor has spectacular views over the Clare coast and backs onto the Cliffs of Moher. These aspects of remoteness and imbrication within natural landscape settings are an essential part of the therapeutic power of place (Kearns and Gesler, 1998). In the wells themselves, the settings, such as the deep dark well at Faughart, surmounted by a large igloo-like stone cone, reflect natural assemblages of water, stone and bush, where light and shade, leaf and branch, path and grass have affective, even pagan antecedents (Strang, 2004). The presence of specific and symbolic trees such as whitethorn, hazel, ash, holly and rowan act as framing elements around well sites (Logan, 1972; Simon, 2000) Almost all of the wells, with the exception of Struell, have rag trees or bushes, with variable volumes of left offering. They are particularly intense at Tobernalt, Faughart Well and Tully Garden. At Liscannor, the ledge by the entrance passage-way provides a metaphoric equivalent (see Figure 2.2). At Doon, older photos of the well show no bush but rather, the remarkable sight of a collection of crutches embedded in the ground in a cluster, on which an intense array of rags and cloths are wrapped and hung (see Figure 2.6). In this setting, the (crucially) discarded crutch, global signifier of the successful cure across both pilgrimage sites and spas, acts simultaneously as a pseudo-rag bush and as therapeutic object, as a signifier of the spiritual power of healing in place.

Figure 2.6 Doon Well, Co. Donegal, c1910
Source: Courtesy of the National Library of Ireland.

When considering the ways in which holy wells are accessed and used and how the places themselves are organized, at all times there are interplays of movement, mobility and stillness. This mobility is acted out in time, but also in place. Wells have moved their physical position, as at Liscannor and Faughart, while the supply of the core 'product', the holy water, can also be ephemeral and intermittent as wells spring, flood, dry up or are blocked as at Doon, Tully and Struell. The offerings and rags left at wells are also ephemeral in their forms, intentions and deteriorations. The utilization of the holy well, while most fully expressed in ritual and symbolic aspects of the pattern also has a mobile everyday and personal expression. In visiting each of the wells several times in a three year period from 2007 to 2009, at no point did I find an 'empty' well, that is, empty of the presence of at least one other visitor. Indeed rarely were there less than five visitors in any half hour period. It is in these slow steady trickles of daily adherents, from young to old, that the wells retain their powers as arguably everyday therapeutic sites.

Throughout their existences, holy wells have also been strongly associated with informal folk medicine practices. In those folk understandings, the healing power of the well was seen as mobile in terms of its seasonality and temporality. Seasonality was embedded in the importance of certain dates for patterns and festivals. Tied strongly to the shape of the agricultural year, many patterns were associated with pagan festivals. These included; *Imbolc* (1 February), the beginning of spring;

Bealtaine (1 May), the beginning of summer; *Lúnasa* (1 August) the beginning of autumn and; *Samhain* (1 November), marker of both end and beginning, as the productive life of the earth stops and winter approaches (MacNeill, 1982). Yet patterns were also moved, as at Liscannor, from St. Brigid's Day, 1 February to Garland Sunday, as pragmatic responses to seasonal weather. Connections to health beliefs were also mobile between pagan and Christian worlds. Illnesses were historically linked to nature; the divine or malediction and these were in turn treated at the holy well by cures which were similarly constructed as natural, divine or counter-maledictive. The waters of the holy well were used in all three of these ways. There has arguably been a revival in the first two of those understandings, through an interest in new age spiritualism and CAM that continues to place holy wells as centres of health and wellbeing (Rickard and O'Callaghan, 2001).

Mobility was additionally seen in particular temporal associations with the healing power of the well (Bord and Bord, 1975; Bourke, 2001). Like energy, the power of the cure was reputed to be an oscillating force that reached a peak on a particular point in time, the stroke of midnight on the eve of the Saint's day (Castlekeeran). This accounted for the timing of the dates though there were also variations on this theme, whereby the peak curative power ran from midnight to midnight on the pattern day, or built up and fell away in the week(s) before and after. One could access the healing power of the well at other times; it just wasn't as strong. These associations of time, place and healing potential were an unusual feature of the holy well, though also found in formal pilgrimage sites such as Santiago de Compostella and the rotating locations of the *Kumbha Mela* in India (Singh, 2006). In addition a conceptualization of the healing power of a well as having both a seasonal (natural) and temporal (energy) component, fits into new age spiritual understandings of place (Shaw and Francis, 2008). While these 'user groups' are still small in relation to more conventional users, the interest of such groups, often affluent and passionate, have helped in a revitalization of interest in wells such as at Tully, Castlekeeran and Liscannor. In drawing attention to how the cultural construction of healing places over time are still strongly shaped by their geographies, the holy well acts as yet another marker of health/place connectivities (Kearns and Gesler, 1998; Gesler, 2003; Williams, 2007). Finally, the therapeutic value of place is also expressed in its very lack of mobility and the phenomenological values of the sites as affective spaces of contemplation, stasis and stillness (Conradson, 2007).

Economic Performances of Health: Holy Water and Souvenirs

In considering economic performances of health at the holy well, the connections might not at first be apparent. Unlike the spa or sea-resort, entrepreneurship was not central to place creation at the holy well (Walton, 1983; Kelly, 2009). Yet in considering holy wells as places of spiritual healing and pilgrimage, their identities as sites of early health tourism becomes apparent (Logan, 1972; Harbison, 1991). While many of the holy wells were in isolated rural settings, especially Doon, Struell and Castlekeeran, and as such had little economic impact on their surroundings, a number of the holy wells were located in or near towns, like Tobernalt and Tully where some indirect benefits accrued. In particular, St. Declan's Well at Ardmore had a history as a pilgrimage destination, one of several sites of devotional and antiquarian interest, all of which featured in pattern day celebrations (see Box 2.3). In addition, both well and monastic site were set on a hill overlooking a pretty seaside village. With both pilgrimage and pattern attracting large numbers, many of the visitors were inclined to linger for a few days (Lincoln, 2000). Here an economic trickle-down effect was expressed in a demand for lodgings associated with holy well visits, becoming in effect a prototype for later seaside health economies (Towner, 1996; Lincoln, 2000). Evidence of local provision of lodgings, both paid and unpaid, was also found at Castlekeeran and Liscannor. These direct and indirect benefits for local economies around holy wells were particularly welcome in a country with observable rural poverty well into the twentieth century, though a tradition of free and reciprocal care should also be noted (Duffy, 2007).

The subtle and evolving commercial identities of religious sites are a common theme in the literature of tourism geographies (Urry, 2002; Timothy and Olsen, 2006a). Historically there were a range of commercial activities associated with pattern days. The presence of tents and stalls serving food and drink was noted at pattern days in Castlekeeran, Liscannor, Ardmore and Tobernalt. In Ardmore, when people came on pattern visits, they brought both food to eat and food to sell (Ó'Cadhla, 2002). As with any social or cultural activity that draws a crowd, there was additional money to be made in other forms too. Other products included religious medals, statues, rosary beads, Mass cards and a variety of votive goods, reflecting the more intensive commercial excesses found at contemporary pilgrimage sites like Lourdes, Fatima or Medugorje (Taylor, 1995). The size of the Irish well and its 'market base' militated against such large-scale development, though stalls selling religious objects were historically recorded at Tobernalt and Liscannor. In a contemporary setting, Lady's Island has a shop beside its church selling a variety of these products, which opens for the three weeks of 'the pilgrimage season' (15 August – 8 September), while 'ephemeral' sellers in mobile trailers also appear at other wells like Tobernalt on pattern days. At almost all of the sites, donation boxes are evident for the collection of monies to support the maintenance of the well. These are most prominent at the bigger sites such as Faughart Shrine, Tobernalt and Lady's Island, reflective of a more voluminous user base. At Father Moore's well the money is collected up and donated to various charities as well as for the upkeep of the well. At some of the sites, including Liscannor and Ardmore, small amounts of money were traditionally given to 'guardians' of the well, whether to the caretaker or an 'old crone', as at Liscannor, or to a woman at Ardmore: 'a formidable figure (who) had possession of it (the well) and dealt out (water) in pint mugs to those who paid' (Croker, 1824; Lynd, 1998). In general however, especially in more recent years, the core 'product' of the well, the holy water, has been freely available for consumption on site and to be taken away in bottles for use elsewhere. Though some of the wells are less hygienic than modern sensibilities might like, this metaphoric quality of water as a potable and portable product is a characteristic of this and other healing places.

Given the role of culture in creating the holy wells, a consideration of cultural tourism has rarely been applied specifically to a holy well (Smith, 2003). While the well itself cannot be sold, it can most certainly be commodified in material and metaphorical ways. A number of statutory and voluntary agencies have become engaged with the preservation and maintenance of holy wells in recent times, exemplified by the Northern Ireland Environment Agency's (NIEA) ownership of Struell. In addition, as was noticeable at Doon and to a lesser extent at Tully and Faughart, there were funding grants available to develop the sites. While a level of genuine local pride and identity is manifest at wells, and certainly local voluntary and parish committees manage wells in that altruistic way, the involvement of local and national tourism organizations may be more contentious. In a sense, holy wells are being rediscovered, or more specifically, appropriated as objects via the more commercial discourses of cultural tourism and heritage (Strang, 2008). Some of this marketing of place alludes to their mysterious and mythical Celtic pasts, as well as the more 'folksy' heritage and kitschness dimensions. This is exemplified by the way in which St. Brigid's Well at Liscannor has been commodified through its proximity to the Cliffs of Moher and incorporation into the 'Cliffs Experience' (see Box 2.1). Local accounts of a notable increase in tour bus visits since 2007 provides evidence of a spatial annexation of a healing place to a more constructed and commercial external 'ownership'. While some might welcome such a development, in encouraging tourists and locals alike to begin to explore other wells and their spiritual and curative values, this may be an optimistic reading of a potentially more negative impact on well sites as contemplative therapeutic settings.

Thinking about holy wells in this way gives rise to a deeper consideration of management, ownership and power (Dorn and Laws, 1994; Kearns and Moon, 2002). Any discussion on the holy well must therefore consider the power structures, which have acted on them over time, and how these manage, as well as commodify, health performances at the well. A wide range of agents of structure have visibly (and often invisibly) shaped the holy well. These included at national level, a range of governmental agencies from church to colonial government to the newly independent state after 1922. At a more local level, local government (county councils), landowners of various types, the church, community organizations and private individuals have had a stake in and an ownership of the holy well. Finally, competition was involved in the different ways that the therapeutic resources (as product) of the well and their associated social and cultural performances have been accessed and used. As the functional constituency of the holy well, the public, whether individuals, groups or geographical collectives, all have an ownership of the well and that ownership inevitably comes up against and conflicts with wider agencies of structure and power.

The newer wells such as at Doon, Tully, Fr. Moore's and especially Faughart Shrine, all have a more regulatory and official church feel, with statues and grottoes, altars, shrines and oratories and even the use of celestial Marian blue on the walls at Doon (Rickard and O'Callaghan, 2001). The oratory at Faughart Shrine and the wider use of statuary and symbolic elements, such as the fourteen Stations of the Cross, also form part of the more ecclesiastical furniture of the place. Yet at a number of sites, there is, within the hybrid mixing of old and new built elements, suggestions of contestation. At Castlekeeran, the lower part of the site is primarily natural. At the top of the hill at the rear of the site sits an elegant stone oratory, a copy of St. Colmcille's house in nearby Kells, built in the 1910s by the local parish priest. While the architectural nod to a monastic past provides an attractive aesthetic counterpoint to the rest of the well site, its positionality and construction present a distinct contrast, replicated at Tobernalt. The symbolic function of the newer built elements of oratory, altar, statuary and grotto mark the imposition of church power and surveillance (from a height) over the more natural and pagan elements of tree and stone at the lower well. In this they may also represent a cultural and moral hegemony of landscape-as-gaze over landscape-as-lived (Taylor, 1995; Wylie, 2007). Yet some newer Marian shrines from the 1950s are now assumed to be wells or have a well identity attached to them via the depositing of left objects so these processes are not entirely one way.

There is also the deeper aspect to ownership in the form of control, which has been enacted at holy wells over time and up to the present day. While the church has maintained a relatively ambivalent attitude to holy wells, it has stepped in at various stages to assert this control. The early church had concerns over the pagan traces of hydrolatry, while the early modern church had issues with the more profane aspects of the pattern (Strang, 2004). For the colonial government, the interplay of violence and group assembly made the pattern the focus for public order clampdowns (MacNeill, 1982). In the mid-nineteenth century, the most concentrated church attempts to control patterns took place and they were banned in a number of dioceses, which affected Struell, Tully and Faughart Well. This very visible control marked a shift from a church as symbol of resistance during the period of the Penal Laws to a more co-operative role with the colonial government in return for the freedom to practice proto-colonial agendas of its own. While the collaborative clampdown on patterns was nominally aimed at spiritual control and social order, there were other contributory factors (Carroll, 1999). An equally important aspect of the clampdown on patterns was the opportunity it provided the church to begin to reshape material and metaphorical meanings at the wells and Faughart is a good example of this process (Bourke, 2001). Here, the original Well, hidden and enclosed in a medieval stone hut on the top of a windy hill with its superstitious metaphors of rag trees, skulls and folk medicine, was re-produced and in a very literal sense, re-placed, to a redeveloped site a mile down the road. At the Shrine, the semiotic and metaphoric uses of Brigidine imagery were given full rein, along with a cast of other church approved saints, material objects and conventional church items, set into a mostly open and visible site. In part this represented a shift from a customary lived curative site, to a modernized landscape shaped by surveillance, in Foucauldian terms, as 'religious gaze' and via the suppression of the performative 'lower' bodies of the carnivalesque (Foucault, 1977; Ó'Cadhla, 2002). In addition, other re-shapings of holy well sites at Tully, Liscannor and Tobernalt suggest an imposition of this religious gaze, keeping the pagan, individual, profane, embodied, free and often female expressions of healing and worship in its (i.e. the Church's) place (Condren, 2002).

While the settings of holy wells are primarily public spaces, some are still in private hands or on private land. While distinctions between ownership and management are important, they are hard to unpick in the often nebulous geographies of the holy well. At Castlekeeran, the land is owned by a local farmer but given over to the parish to manage and run in effective perpetuity. Local committees at Faughart, Fr. Moore's, Liscannor and Tully are all actively engaged in managing the sites in conjunction with the local parish and local authorities. This demarcation is more clear-cut at Tobernalt, where the parish owns the site. In 2008 this was expressed in on-going renovations and additions to the site, funded from visitor's offerings and the sale of books on the well. At a number of the sites, the local county council has a very clear role as manager and shaper of the surrounds of the well. At Ardmore, Doon, Lady's Island and Tully, there are symbols of county council involvement in the form of road-signs, road management and public facilities such as toilets and car-parks. At Doon, in particular, the well has received a considerable make-over in the past decades, a process funded in part by Donegal County Council but also, rather surprisingly, by the European Union under the auspices of grants from a Peace and Reconciliation Fund.

While wells were commodified directly through the occasional sale of religious objects, it could be argued that their economic performance is more subtly expressed through two other forms of capital, spiritual and cultural. Given the limited resources, the riches at holy wells, such as they were, were expressed more as a conceptual ownership and re-harnessing of their spiritual power to the needs of the church and state. There were no great financial benefits but rather a moral geography and economy of citizenship and reverence (Taylor, 2007). It was not so much about the capture of pockets as the capture of hearts and minds (by the state) and souls (by the church). In recent years, as church and state roles have lessened considerably, newer conceptualization of control, through a recognition of the well's potential as cultural tourism and heritage sites have re-emerged. This selling of place has yet to be contested in the public arena against more impartial meanings linked to the spiritual and popular meaning of the holy well as therapeutic landscape. More importantly those affective, behavioural and performative aspects of individual ownerships of health and wellness in place need to be considered. Restoration, memory, healing, reflection and the consumption of the holy water would be more difficult to sustain at a site unloading coach loads of tourists whose vision of the well, though acceptable in its own right, may be very much at odds with the therapeutic dimensions of those places (Conradson, 2005b). In trying to link mind-body-spirit experiences in place with the cultural productions and controls (of hearts, minds and souls) that also shape those places, ongoing contestations of health meanings at the well remain evident.

Summary: Holy and Holistic

> About three perches to the right is the biggest of its wells, it is called the Holy Well. At one side there are three steps leading to the water. A metal cup hangs by a chain on a past. People drink from this. Around the well are placed 5 blank iron crosses about 2 feet high. By going around the five crosses three times, the pilgrim makes what is called a 'station'. People would not use the water in this well for any purpose other than drinking. Eileen Coulds (Aged 14) (NFC, 1938, 159)

Between 1934 and 1938 the still new Irish Free State government set out to collect information on holy wells, which included the cures associated with them, the timing of pattern days as well as folk-cures used more widely in the surrounding area (NFC, 1934a; 1934b; 1934c). These collections provide much of the evidence we have on the healing powers and performances at specific wells and were written up (under supervision) as free text and drawings by local school children (including my own mother). From this we have multiple accounts of the specific cures associated with individual wells, as for St. Kieran's above, which though contestable in their academic accuracy, are nonetheless valuable in being a rare record of more embodied local voices, or what Lorimer refers to as 'small stories' (O'Reilly, 1998; Ó'Cadhla, 2002; Lorimer, 2003).

In visits to St. Kieran's Well (see Box 2.2), the rituals and performances encountered reflect centuries-old narratives of water, health and place rooted in a belief in the power of the well as a physical faith cure. In discussing how therapeutic watering-place identities were created within material settings, the importance of narrative, metaphor and discourse cannot be underestimated. In the same way that Boorstin referred to the notion of the 'pseudo-event' in tourist practice, that same metaphor can be used in relation to the 'pseudo-cure', where arguably 'inauthentic' practices become real through repetition (Boorstin, 1964). While this notion is more fully explored in other settings around contested health narratives in place, the methodological importance of myths, symbols and even superstitions (in Irish, *pisreog*) were essential in the production of the hydrotherapeutic landscape.

The presence of the holy well as a public and free site of care and as a watering-place for the masses marks it out from some of the other forms of therapeutic landscape discussed elsewhere in the book. It is not too fanciful to see the holy well as the 'people's spa', echoing Canadian First Nation descriptions of nature as 'a 24-hour pharmacy' (Wilson, 2003). It was used by rich and poor alike and has functioned for over a thousand years as a primary site for medical attention. Along with the use of skilled formal and informal medical healers, wells formed part of a proto health care system of their day. In addition, they represented the discursively created medical beliefs of their time, through the invocation of god and nature. As understandings of health became more medicalized, rational and scientific, those more spiritual and natural aspects of health became less celebrated. There is also a suggestion that the decline of holy well use was also a psychological and pragmatic process. The impact of the Famine was said to have lessened faith in God, while the guilt-ridden colonial government introduced more formal systems of health and social care, such as a national network of dispensaries (Carroll, 1999). Along with the rise of the country doctor, these alternatives to the holy wells were eagerly adapted, supplying an evidential rather than belief-based cure. Nonetheless there has been a revival of interest in holy wells as sites of healing as dissatisfaction with conventional biomedicine has emerged in the past decades (Logan, 1981). This has led to a greater focus on CAM that incorporates folk medicines and traditional cures, which coupled with the rise of new age spiritualism, has led to a renewed reputation for the holy well as site of healing and wellness (Bourke, 2001).

In considering holy wells as both therapeutic landscapes and sites of therapeutic experience, it is also important not to neglect the places themselves. While acknowledging the complexity of the space/place debate and the different understandings that can be applied to place, holy wells take particular physical forms which are concrete, affective and lived (Kearns and Gesler, 1998; Thrift, 2008). This is evident in those specifically therapeutic aspects of place, such as the affective settings of the well sites as natural places of peace, restfulness and contemplation, constructed around grass, tree, stone and running/still water. Here, the psycho-cultural links between wells and wellness are further strengthened. The forms of health which lie at the heart of the holy well are spiritual in their foundations, yet also speak to wider holistic understandings of healing and wellbeing. St. Declan's is known as a *lóc fás*, a liminal space with retreat/hermetic space identities, translated locally as a hermitage (Ó'Cadhla, 2002). Such perspectives would certainly fail to convince proponents of a vision of health that is more focused on disease, illness and scientific cure, and who would not have much sympathy for a spiritual or faith perspective either. Yet it is not the intent of this work to reverse such an opinion. In the historic and ongoing contemporary inhabitations of the holy well, narratives of healing, peace and cure speak for themselves as complementary curative experiences.

The role of embodiment is essential to understanding human engagements with the well as healer. From the physical acts of drinking the holy water and rubbing it on the body, to the removal of diseases in symbolic but embodied forms via rag tree rituals, all represent a form of healing that is at heart phenomenological. It is lived, experienced and enacted in place. The body, through external contact with the earth, and internal contact via the ingestion of water (and as an affective felt site of cure), is central to that embodied relationship with health in place. In addition, the material and metaphorical connections at the holy well between health, place and the body, are further deepened by individual and communal performances and ownerships. One can see in the rounding rituals and the non-representational performances of health a visible example of what Lorimer calls: '...

shared experiences, everyday routines, fleeting encounters, embodied movements, precognitive triggers ... affective intensities' (Lorimer, 2005, 84). Those affective intensities are experienced in places which in their natural and built therapeutic designs trigger such emotional responses. They are also experienced primarily through embodied chronic and even psychosomatic illnesses that orthodox medicine finds difficult to deal with (Philo and Parr, 2003). The performances of health in place can also be theorized as a theatrical performativity, a quite literal enactment of health in a symbolic set(ting), to a prescribed but interpretable text, with an unseen director and a cast of strolling players who are part-actor, part-audience (Shackley, 2001).

The identity of the well as a notional 'people's spa' and the people, native, clerical and colonial, who shaped that identity, deserve deeper consideration. In keeping with a more grounded political economy perspective, holy wells also represented deeper currents of structure, agency and power. A number of governmental and private agencies were and are engaged in the production/reproduction of place with wider agendas linked to political and moral control and the commodification of the sites distinctive cultural heritage potentials. At Liscannor, Healy (2001) saw this as a contestation of therapy and tourism. In all of these structural shifts, the original therapeutic meaning of the holy well is tested but continues to resist through individual and communal acts sustaining that original meaning. It is these psycho-cultural acts of ownership which provide the counter and balance to the wider cultural forces acting on the holy well as therapeutic place. In considering the historical ownerships of bodies at the well, the notion of a moral geography has value, wherein enactments/performances and self-place experiences are always framed by wider structures and moral judgments by priests and medics, ironically the joint function of the pagan shaman (Knott and Franks, 2007). The negotiation of an often contingent ownership of health in such places also mirrored wider contestations of health paradigms between the orthodox and alternative. Taking this forward, the next setting of healing waters to be discussed is the historic spa town. The connection between the holy well and spa should be noted in two ways. Much confusion arose in early modern descriptions of Ireland between the holy well and the spa, so that there are sufficient overlaps to justify suggestions that the holy well represented an earlier form of spa (Carroll, 1999; Kelly, 2009). The decline of the holy well was additionally linked to the development of a more formal spa culture in Ireland from the mid seventeenth century on. This was shaped by the identity and culture of the new ruling colonial classes, whose position in a more aggressively settled post-Reformation Ireland showed little interest in either 'Popery' or the peasantry and so needed to find new secularized healing waters of their own.

Chapter 3
Spa Towns: The Rest Cure

Introduction: *Sanitas Per Acqua*

> … our worthy hero prepared for his journey to this once celebrated Spa, which possessed even then a certain local celebrity that subsequently widened to an ampler range. The little village was filled with invalids of all classes; and even the farmers' houses in the vicinity were occupied with individuals in quest of health. The society at Ballyspellan was, as the society in such places usually is, very much mixed and heterogeneous. Many gentry were there gentlemen attempting to repair constitutions broken down by dissipation and profligacy; and ladies afflicted with a disease peculiar, in those days, to both sexes, called the spleen a malady which, under that name, has long since disappeared, and is now known by the title of nervous affection. There was a large public room, in imitation of the more celebrated English watering-places, where the more respectable portion of the company met and became acquainted, and where, also, balls and dinners were occasionally held. (Carleton, 1860, 434-5)

Driving on the main road from Dublin to Cork, one passes through the village of Johnstown (Co. Kilkenny) which contains a small but perfectly-formed octagonal central square. If you glance at a house on the north-west side, you see its name, 'The Spa House', writ large on the front wall. Though the town's historical reputation as a major health-resort has long passed, especially as it was then known as Ballyspellan, the lingering palimpsest of name and function may pique the curious visitor's interest. In a rare Irish example of an imaginative account of a spa town, Carleton's comic novella of 1860, *The Evil Eye* or *The Black Spectre,* sets part of the story in Ballyspellan. The text describes a town based on the typical British watering-place template but with unique inhabited aspects of its own, where dissipation and disease were equally prominent, a pattern repeated in performances of health across the Irish spas.

The spa town has had a particular attraction for health geographers given its unique position as a place developed explicitly around a health metaphor (Gesler, 1993 and 2003). Yet outside of studies of Bath and Hot Springs (ND), there is surprising little detailed research by health geographers on spa towns (Gesler, 1998; Geores, 1998; Valenza, 2000). This gap is filled primarily by historians and tourism geographers who have provided a range of studies at local and national levels, covering the British Isles, New Zealand, France and North America (Wightman and Wall, 1985; Rockel, 1986; Hembry, 1997; Mackaman, 1998; Durie, 2003a). Other countries with strong spa traditions, such as Germany,

Hungary, Czech Republic, Korea and Japan also have extensive literatures in their respective native languages (Hahn and Schönfels, 1986; Smith and Puczko, 2009). In addition medical historians continue to develop the subject, albeit with a focus on traditional understandings of medicine/health (Brockliss, 1990; Kelly, 2009). While many of these studies provide a sound basis for studying the cultural and economic development of the spa towns, few have explicit theoretical concerns with therapeutic aspects of place (Kearns and Moon, 2002; Williams, 2007). In particular, different aspects of performance in place seem ripe for exploration within the culturally produced regimens of spa town society (Lorimer, 2005; Rose and Wylie, 2006).

Typically, spa towns developed around mineral springs, hot or cold, with a reputation for healing (Gesler, 1992). While that development in the early modern period was associated with the town of Spa in modern Belgium, spas had much older roots in Greek and Roman cultures (Hembry, 1990; Jackson, 1990). Their development varied depending on a range of factors (Porter, 1990; Durie, 2003b). Historical notions of development cycles help explain why some spa towns developed and sustained while others barely got off the ground (Hembry, 1997; Kelly, 2009). Entrepreneurship and competition linked to health/place promotion were also central to their development, aided by crucial though unpredictable aspects of reputation and fashion (Gesler, 1998; Kelly, 2009). All were commercially developed as 'going concerns' and some, like Rotorua or Vichy, were specifically constructed to compete with spas in other jurisdictions (Rockel, 1986; Mackaman, 1998). Spatial elements such as water chemistry, accessibility to population centres, improved transport and wealthy clienteles were also important in the successful production of place (Palmer, 1990; Gesler, 2003). Finally, spa towns were historically associated with spring to autumn seasons and daily routines of medicinal consumption, exercise and diet, in addition to less healthy leisure and pleasure pursuits (Brockliss, 1990; Gesler, 1992).

Within this wider context, the position of Ireland is, at first glance, a limited and peripheral one. Unlike many British and Continental spas, there were no hot springs apart from Mallow and the spa there was at best, tepid at 70°F (Myers, 1984). Within Ireland, Kelly's medical history identified a spa town development which was limited in terms of its social and cultural impacts (Kelly, 2009). A view of Ireland as peripheral and limited was also indicative of Ireland's ascribed contextual position as simultaneously a colony and part of Britain for almost 800 years (Duffy, 2007). By the late seventeenth century, that colonialism had become more embedded and it was into such a setting that the imported spa town form was introduced. It was at its peak from the late-seventeenth to the mid-nineteenth century, after which competition from sea-bathing and a general decrease in interest in the watering place as curative site led to decline; though the advent of hydrotherapy provided a small revival between 1850 and the 1930s (Kelly, 2009). As therapeutic landscapes, the Irish spa towns reflected core settings identified by Gesler (1992). Built and natural environments were exemplified by spa specific elements such as baths and pump rooms alongside landscaped walks and hotels.

Symbolic environments were represented in discourses of healing waters, their chemical and curative powers, and generic narratives which promoted metaphors of spa=health (Geores, 1998). Finally the social environments represented in the facilities, hierarchies, clienteles and cultural practices of the spa were as evident in Ireland as in any other part of the colonial world (Urry, 2002).

The specific mineral waters varied considerably and seven different categories of spa waters were identified (Rutty 1757). The two most common forms were Chalybeate (Iron) and Sulphur spas while others included magnesium, saline, alkali and vitriolic. These classifications were primarily linked to the dominant minerals in the waters and the extent to which the waters retained their healing powers over time and space. The spas/springs were found in most parts of the island, with approximately 200 different spas or springs listed from a variety of sources (Rutty, 1757; Ryan, 1824; Lewis, 1837; Knox, 1845). As was the case in England and Scotland, only a very small number of these sites developed their infrastructures such that a town subsequently grew around them (Hembry, 1990; Towner, 1996). While the colonial setting of the Irish spa always looked first to Britain, there were also less well-documented associations with continental spas like Spa or Carlsbad (Karlovy Vary) which provided a more nuanced understanding of proto-globalization and competitive health markets found in even the most liminal of locations (Kelly, 2009).

Previous studies of spa towns have focused on material and metaphorical aspects of their production as healing places and even as 'fountains of youth' (Valenza, 2000; Cossick and Guillou, 2006). Employing a therapeutic landscape approach may help to re-place inhabitations of health more centrally into those narratives (Kearns and Gesler, 1998; Gesler, 2003). As early sites of health tourism, the production of the spa town's reputation was mostly discursive and sold on the chemical and scientific studies of their waters and specific illnesses to which that water could be applied. These were produced by medics and local entrepreneurs, and consumed by an eager patient clientele (Durie, 2003a). More importantly, the practices and performances of the cure in the spa towns of Bath, Mallow or Vichy, were essentially cultural productions; where social elites, aspiring practitioners, eligible women and even beggars were attracted to both the healthy and unhealthy identities of the towns (Porter, 1990; Mackaman, 1998). These cultural productions were subject to shifts in their reputations which at times aligned with parallel fashions of health. In nineteenth century French and English spas, newer and more medicalized forms of cold water treatment associated with hydrotherapy became popular, a process repeated at Lisdoonvarna and Lucan (Price, 1981; Mackaman, 1998). A number of towns like Bath and Spa sustained a consistent popularity over time, emphasizing the metaphoric power of the word spa and its equivalents.

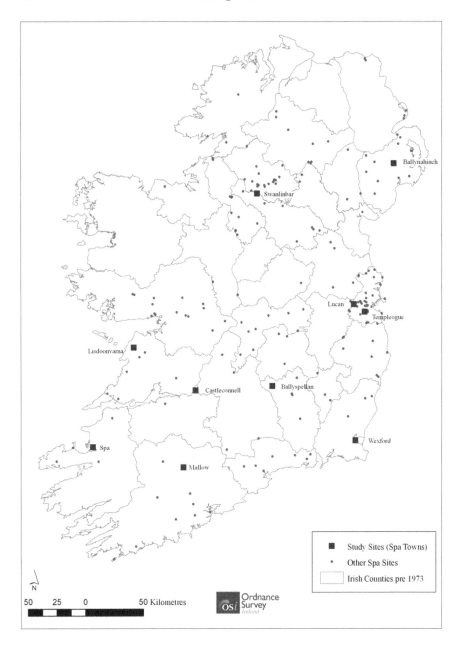

Figure 3.1 Map of the Distributions of Spas with 10 place-study locations
Source: Rutty, 1757; Lewis, 1837; Henchy, 1958; Boundary Maps, Ordnance Survey,
Ireland. © Ordnance Survey Ireland/Government of Ireland Copyright Permit No. MP
008009.

The rest of the chapter explores health and wellness at the spa through a set of embodied performances linked to wider cultural discourses creating and sustaining those places as well as the deeper (per)formative roles of space and the economy. As in the previous chapter, three specific towns: Mallow (Co. Cork: see Box 3.1), Lucan (Co. Dublin: see Box 3.2) and Lisdoonvarna (Co. Clare: see Box 3.3) are described in detail while the remaining place-studies are drawn from the towns of: Ballyspellan (Johnstown) (Co. Kilkenny), Swanlinbar or Swadlingbar (Co. Cavan), Ballynahinch (Village of Spa) (Co. Down), Castleconnell (Co. Limerick), Spa or Tralee (Co. Kerry), Templeogue (Co. Dublin) and Wexford (Co. Wexford). All of these towns feature in Kelly's recent exploration of the medical history of Irish spas for the years 1680 to 1850, within which he also notes the significance of the towns across different periods of operational and reputational fame (Kelly, 2009). Other smaller towns and locations functioned as 'local' spas and are noted where relevant (see Figure 3.1).

Embodied Performances of Health: Sulphur and Scurvey

An interest in the study of Irish spas can be traced back to the short-lived Physico-Historical Society (1744–52), one of whose members, John Rutty subsequently produced a definitive book on the mineral waters of Ireland (Rutty, 1757; Magennis, 2002). Containing full and dull descriptions of a range of chemical tests, the book also provided long lists of cures encountered at 128 different spas. While written with all the chemical and medical brio it could muster, the book was at heart, as with spa guides produced in other countries, a promotional document (Granville, 1971; Knox, 1845). Explicitly noted by Rutty in his introduction, was an attempt to persuade the Anglo-Irish of the quality of the product in their own backyard:

> It is true, we cannot boast of hot Baths, but with respect to other medicinal waters which have amply recommended themselves by their good effects, we vie with our neighbours: and first as to Chalybeate waters, we have perhaps as great a plenty and variety as any country of equal extent in Europe; and some of these so strongly saturated that they might in a great measure supply the place of the German Spa water. (Rutty, 1757, ii)

Thus from the outset there was a strong discourse promoting local spas aimed at creating a domestic market to tap in to a demand then being supplied from both England and Continental Europe (Rutty, 1757; Knox, 1845). In this they were no different to early forms of tourist guide-books or catalogues which was why a full listing of chemical tests and documented cures were essential components (Coley, 1990; Hamlin, 1990; Horgan, 2002).

In considering embodied performance of health at the Irish spa, the list of conditions for which cures were to be found was lengthy and covered virtually all known illnesses of the time. While infectious illnesses were generally discouraged, cures for conditions like leprosy and dysentery were noted at Ballyspellan and Swanlinbar respectively. More typically, each spa had its own 'menu' of conditions for which the waters were proven curatives, through 'insider' chemical and medical reports and of courses patient narratives, brought back or reported to homes via letters (O'Sullivan, 1977). Gout was treatable at almost all of the representative wells, especially at Swanlinbar, Wexford and Lucan while 'barreness' was treatable at Castleconnell, Wexford and Ballyspellan. Even a small village such as (Tralee) Spa had recorded cures in the 1750s for: colic, rheumatism, scurvey, liver complaints, dropsy, phlebitis, pleurisy, agues, scrofula, cachexies and hysteria. The lists of diseases changed over time as new terminologies and ironically, illnesses became popular. While the fashionability of place is regularly cited in social and medical histories, it was replicated in a linked 'fashionability of illness' (Porter, 1999). Newer named conditions such as psoriasis, bronchitis and dyspesia were treated at the later spas of Ballynahinch and Lisdoonvarna. In addition, gender played a role, with different treatments listed for a range of ascribed female complaints including; obstructions, weaknesses, reproductive problems, flushings and the old reliable, nervous hysteria (Rose, 1993; Bergoldt, 2006). In this the spas provided early evidence for the public display and discussion of women's health issues, even if those discussions reflected the flawed attitudes of the time. Ryan also listed cures for a range of mental health conditions such as depression and phobias and traced a connection between the presence of such cures at the spa and the holy well (Ryan, 1824).

Spa towns were both figurative and literal landscapes of health consumption, and the benefits had to be visible in both bodies and place, to promote and sustain the town's healing reputation (Gesler and Kearns, 2002). Almost all spa town narratives typically followed a process whereby a spring or water source was discovered. This was then popularized, or more specifically, promoted by local interests. A key step in that process was procuring an initial chemical analysis from an esteemed scientist. This was followed by a medical account of cures to be found at the place, preferably produced by a reputable physician. While there were some examples of contestation between chemists and physicians, the evidence of the doctor was considered more important in the long term, in that the physical properties of the water could be directly linked to its specific medical benefits (Coley, 1990; Hamlin, 1990). Thus for example, the identification of a large amount of iron in chalybeates was discursively converted into medical parlance and deemed good for cholic and ague at Ballyspellan (Rutty, 1757). But this medico-chemical evidence varied from spa to spa and even from spring to spring within spas. It was also used in a 'competitive health promotion', whereby the benefits of individual waters, based primarily on their chemical contents, were compared, usually favourably, to the 'inferior' product obtainable elsewhere in Ireland, Britain or the Continent (Hamlin, 1990). The waters at Mallow for example, were regularly noted in local promotional material to be the superior of the Hotwell at Bristol (Knox, 1845). This process of bad-mouthing other spas was also common practice in Britain and France (Brockliss, 1990;

Borsay, 2000). Mackaman noted that ambitious nineteenth century French spas, operating in an increasingly medicalized milieu, had a tendency to describe the curative properties, or more specifically the limitations, of every spa but their own (Mackaman, 1998). While this tactic was less explicitly used in Ireland, collective narratives of bodily cures and relational reputation remained central to 'health-making' at the spa.

When it came to the embodied practices of taking the waters, the two primary forms, common across watering-places, were ingestion and immersion; at least in the period up to the mid-nineteenth century. At the spa, the phenomenological relationship between person and curative object, the mineral- and by extension, health-enhanced waters, was fundamentally embodied. In drinking the water, the 'medicine' was introduced into and flowed through the body, while bathing in the water was an even more embodied engagement involving immersion, a literal entry of the body into the 'medicine'. A range of different baths (salt, perfumed, medicinal and vapour) were recorded for example, in 1829 at Mallow and at Lucan at the end of the century (Jephson, 1964). While recent biomedical opinions are broadly dismissive of such treatments, there is evidence to suggest that the processes of consumption, and especially immersion, in water have a physiological effect on the body, causing positive balancing changes in temperature, pulse rates and general circulation (Harley, 1990; Coccheri et al., 2008). To the spa users of the eighteenth and nineteenth centuries in Ireland, this effect was clear and noticeable in an embodied relationship between the water as curative object and the body as healing subject, exemplifying the notion of the 'therapeutic experience' (Conradson, 2005b).

While the consumption of foul-smelling and even fouler-tasting sulphuric waters was rarely a pleasant experience, and the purgative effects of several pints, even less so, the ultimate effect to the body was of a detoxified and cleared system. The waters at Ballynahinch, a mixed sulphureo-chalybeate well, were said to, 'taste and smell like water that has been used in scouring a foul gun', while Rutty also noted of the same waters:

> It is drank from three pints to four quarts; the chief Operation is by urine, some it vomits (a frequent effect of other sulphureous waters) and it is said to have purged others, but this last effect seems to have been meerly accidental, as the quantity of the Salts it contains is inconsiderable; it does not chill the stomach, like common water, when drank in large quantities, though it occasions sulphureous belches. (Rutty, 1757, 460)

To help 'sugar the pill' in these cases, certain dilutions or additions were noted so that at Castleconnell, the water was taken with milk or, 'If unpalatable it can be eaten with caraway seeds or candied orange peel' (Ferrar, 1767). A frequently cited example from Mallow, and one which was doubtless copied enthusiastically elsewhere, suggested that 'a very little of it as uncommonly good with a lot of whiskey'. Though originally quoted in Rutty (1757), this particular (anti-)curative vignette was invariably repeated in subsequent visitor's accounts, an example of the repeated narrative as metaphor, which when added to Mallow's already rakish reputation, provided a clear discursive connection between health and pleasure.

It was rare to see any account of spa towns which discussed the water as innately dangerous. The fatal disease of PAN (primary amoebic meningoencephalitis) noted at Rotorua and Bath was linked solely to hot springs, a form not found in Ireland (Rockel, 1985; Harley, 1990). Nonetheless there were a number of examples in the guides and traveller's accounts of deaths during or after spa visits. There was however a specifically medicalized rationale for their utilization. Cases of a young child dying after treatment at Ballynahinch and the sudden death of a young woman at Mallow were explained away as due to reasons which had nothing to do with the water (Rutty, 1757; Knox, 1845). In the case of the young woman at Mallow, she was 'discovered' to have had a previously unknown heart condition. However it was the case of the young boy in Ballynahinch which tallied with wider narratives (Harley, 1990; Hembry, 1990). Here the boy failed to properly carry on the treatment once he left the spa and this was the reason he died. Variations on this story in British and French spas, were both cautionary and normative, suggesting crucially, that it was only 'proper medical supervision at all times' which provided the cure and that there were inherent dangers in a failure to adhere to the 'medical gaze' (Harley, 1990; Mackaman, 1998). Knox recounted the need for variations in the volumes to be drunk, noting the mobile relationship between illness and cure at Ballynahinch, with the subtext being the need for the gaze of the regulatory medic:

> The quantity to be taken varies of course with the constitution of the patient and the stage of the disease; and much injury arises in some cases, as I have had opportunities of observing, both from using it, when unsuitable, or in proportions greater than the patient could bear. Half a pint twice a day may be looked on as a proper quantity to begin with, increasing the dose according to its effects or the necessity of the case (Knox, 1845, 268).

The embodied and personalized nature of the cure was such that the performances were shaped at all times by one's own condition, and the extent to which one's material body could cope, both physically and mentally, with the recommended regimens. In keeping with the times, in a medical sense, self-monitoring and self-medication were performed to an embodied response that was in turn tied in to an affective register (Porter, 1992; Holloway, 2006). Dorothea Herbert, a correspondent from Mallow noted that her father, 'moved between illness and health', a realistic account of the ways in which chronic ill-health ebbs and flows (Jephson, 1964).

Box 3.1 Mallow, County Cork

Though the warm water spring at Mallow was originally a holy well, and had a reputation as a spa from the 1680s, it was 'discovered' in 1724 via a specific patient biography documented by Dr. Rogers. The land on which the spring stood was owned by the Jephson family who were central to its subsequent development. Traveller's accounts from 1732 recorded an already active morning clientele at the spa (Kelly, 2009). The peak years of the spa were from the 1730s to 1780s with the development of the Long Room (1738) and the construction of an artificial canal, lined with poplars, which partially routed the spa stream to run under the Assembly Rooms (see Figure 3.2). Its curative reputation and social cachet were sufficient to attract aristocrats and gentry from around the country. The role of the Jephson family declined in the 1780s and the spa house and well were leased out to private management. A small revival was achieved by Charles Denham Orlando Jephson in the building of a mock-Tudor spa-house in 1828. Yet despite the introduction of more hydropathic services including hot, cold and vapour baths, the revival was short lived. In 1857, a new pumped supply of the water was built at the Dog's Heads for public access. Despite the construction of the new Spa Terrace and Road around the same time, competition from sea-bathing and other spas saw Mallow eclipsed (Kelly, 2009). Interest in the thermal waters remains however and the old spa-house is currently the headquarters of a geothermal company.

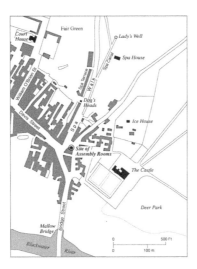

Figure 3.2 Map of Mallow Spa Quarter, Co. Cork, 1875
Source: © Ordnance Survey Ireland/Government of Ireland Copyright Permit No. MP 008009.

The more general therapeutic setting of Mallow, in a 'most favourably circumstanced … medical topography' were heavily promoted in the eighteenth century by a range of local agents. Medical men provided lengthy lists of cures which included a range of generic cures for skin, stomach, phlegmy and urinary conditions (Rutty, 1757). It was also used in cases of nervous hysteria and in later periods for chlorosis, hemmorrhoids and diabetes.

Several more negative accounts, mostly by independent visitors were also circulated, and vigorously contested by the owners/managers of the spa. Metaphorical acts of spatial appropriation, comparing Mallow favourably with Bristol Hotwell and utilizing terms like 'the Irish Bath' helped sustain therapeutic reputations. Additional cultural performances, placing Mallow as 'a resort for the wealth and fashion of Ireland', saw visits from an assortment of aristocracy and the Speaker of the House of Commons in 1763 (Twiss, 1928). The town had a liminal reputation with the activities of wild young men known as the 'Rakes of Mallow', though these were also part of a wider embedding of Anglo-Irish cultural identity. In addition, the attractions of the town, whether in wildness or wellness, were accompanied by a commodification in the form of accommodations, food and drink and transports. One of the noted causes for the failure of Mallow to develop was overcharging. This when added to traveller's accounts of faded grandeur produced in turn a loss of confidence, an essential component of entrepreneurship in place, which led to a spiralling decline in Mallow's social and healing reputation in the 1840s.

The use of metaphor and narrative were at the heart of place-making and the sustenance of a healing reputation (Kearns and Gesler, 1998). Evidence of the effectiveness of the waters (though almost impossible to verify scientifically) was the life-blood of spas as it provided the hard(ish) evidence that a visit was good for your health (Kelly, 2009). Evidence usually took the form of testimonials written by both doctor and patient and this use of 'patient biography', was a favourite tactic in the creation of an embodied connection through narrative between person, place and cure (Rutty, 1757; Mackaman, 1998). A counter to the flow of health/ place promotion did exist and there were plenty of skeptics amongst the boosters. The poet Jonathan Swift notably published a rebuff to a fellow medic, Sheridan's, effusive praise of the cures of Ballyspellan (Henchy, 1958). In Sheridan's original verse the benefits of the water for curing skin problems and dropsy were noted:

> Tho' Pox or Itch, your Skins enrich, With Rubies past the telling,
> 'Twill clear your Skin before you've been, A Month at Ballyspellin
> If dropsy fills you to the Gills, From Chin to Toe tho' swelling,
> Pour in, pour out, you cannot doubt, A Cure at Ballyspellin.

Swift's response, a very clear expression of contested narrative noted:

> Howe'er you bounce, I hear pronounce, Your med'cine is repelling,
> Your Water's mud And sowrs the blood, when drunk at Ballyspellin,
> Those pocky Drabs, To cure their scabs, You thither are compelling,
> Will back be sent, Worse than they went, From nasty Ballyspellin.

Perversely the instant cure was not much favoured as 'return business' (reflecting a tourist identity) was preferred, which at times suited the visitor, given that the social side of the visit was often as important to them as the medical. If you were cured you had no reason to come back, so it suited both parties in a co-dependant relationship to characterize cures as requiring constant and regular annual visits of a month or more. As such there was a clear favouring within spa towns of treatments for the chronic over the acute, though they provided for both. There was even a social cachet associated with requiring a regular cure, which spoke to a certain 'affluential' symbolic connection between health consumption and demonstrable status (Mackaman, 1998).

Over time, the development of ancillary medical treatments in the form of massages, mud baths, electro-therapies and other treatments supplemented the taking of the waters (Porter, 1990). While most of the Irish spa's had passed their prime by the mid-nineteenth century, the two towns that continued to function effectively, Lucan and Lisdoonvarna, reflected the new medical paradigms and treatments associated with hydrotherapy (Kelly, 2009). Prior to 1850, spas had medical treatments which focused solidly on the ingestive and immersive, but the development of hydrotherapy brought with it an unusual combination of embodiment and technology. Developed initially by Priessnitz at the then Austrian spa of Gräffenberg (now Lázně Jeseník, Czech Republic), new treatments using complex contraptions that provided massages, showers and douches, became the new orthodoxy at most spas (Hembry, 1997). Hydrotherapy is discussed more fully in the next chapter, but as Lisdoonvarna was developed from this period, it was inevitable that it would feature these new forms of treatment (see Box 3.3). These were provided in a separate building alongside the pump-room from around 1875 which in time were transferred to a newer building at the rear of the site between 1939 and 1945. The second coming of Lucan (see Box 3.2) was marked by an explicit naming as the 'Lucan Hydropathic and Spa Hotel Company' where a variety of baths and douches were available. These embodied, and at times invasive treatments did however, mark a key phase in the ongoing medicalization of the spa; in turn reflecting shifts in understandings and acceptable health paradigms in society as a whole (Mackaman, 1998).

The spa as therapeutic landscape was initially shaped in its healing metaphors by the minerals of the water and the simple but often effective treatments of drinking and bathing. Over time, and arguably as a corrective to the perceived social excesses, that simple treatment became more medicalized and supervised. But in its recasting of symbolic aspects of health and healing to this more rational 'treatment' based system, the process echoed Weber's notional 'disenchantment of the west' as well as the unstated aims of the spa doctor. This shift away from the 'mysterious realm of water lore', associated with the holy well, was thereby successfully achieved (Mackaman, 1998). More significantly, the practices of self-medication were reduced which impacted on ownership and the more felt experiential aspects of health.

Cultural Performances of Health: Minerals and Matrimony

When considering the spa town, it is often through the lens of an imaginative social geography, based on the romantic adventures of Anne Elliott at Bath in *Persuasion* or the well-fed folk encountered by Humphrey Clinker at 'Harrigate' (Smollett, 1984; Austen, 1993). This connection between people, place and behaviour was embedded in the cultural performances of health at the spa (Gesler, 1998; Williams, 1999c). Indeed for some observers the social life of the spa was its *raison d'etre,* rather than any specific healing role (Hembry, 1990; Kelly, 2009). Whatever the respective merits of the social and curative functions of the spa town, the two were inextricably linked, at least in terms of the metaphors of place identity and performance (Rockel, 1985; Geores, 1998). The social life of the spa would not have emerged without their formative meanings as therapeutic places, while their flowering and survival would not have been sustained without their attractiveness as sites of liminality, leisure and pleasure (Towner, 1996; Gesler, 2003). In this the Irish spa towns mirrored the contradictory settings of holy wells as carnivalesque sites of healing and profane behaviours. While daytime at the spa was given over to the pursuit of health, the evenings were given over to pursuits of a distinctly unhealthier hue, characterized by the description of Lisdoonvarna as 'like Lourdes in the morning and Monte Carlo at night'. At the heart of these activities were the metaphors encountered in the imaginative literature – in themselves distillations of place identities – of food, drink, sport, dancing, gambling and courting.

Box 3.2 Lucan, County Dublin

Given its proximity to Dublin, Lucan had significant potential as a therapeutic place. The town was originally associated with a chalybeate spa but around 1758, an initial phase of development was associated with the discovery of a separate sulphurous source on the banks of the River Liffey. The local landowner, Lord Vesey, was quick to notice its commercial potential and enclosed the well while opening it up for 'business'. It was 'promoted' by a publication of Rutty's which recounted the chemical content and specific cures associated with the spa (Rutty, 1772). By the end of the eighteenth century a new spa hotel had been built along with a terrace of houses known as the Crescent, which were used for visitor accommodation. At this time, the spa was considered a rival to Tunbridge Wells and Leamington. The first phase of the spa ended around 1840, in part due to the increased interest in sea-bathing, but also due to greater social attractions elsewhere (Kelly, 2009). By 1846 the hotel was in use as a school for clergymen. However it had a second coming in 1891 when a group of businessmen created the Lucan Hydropathic and Spa Hotel Company Ltd. It was built on the hill behind the old hotel site with 100 rooms, a range of treatments and a new wing containing the aspirationally named 'National Spa and Hydro', with pumps directly connected to the spring and accessed by an underground tunnel (Madden, 1891). This second phase was aided by a direct tram link to the city and was successful until the 1920's. Lucan is unusual in having a health history covering an earlier chalybeate well, two main development phases, and an intermediate identity as an asylum, for 'Idiotic and Imbecile Children' in the 1870s.

Figure 3.3 Pump Room Interior, Lucan Spa, Co. Dublin, c1895
Source: Reproduced by kind permission of David Cotter Postcard Collection, South Dublin County Libraries.

Reputed treatments at the spa included those for gout, eczema, rheumatism and dyspepsia as well as more generic ailments. The two phases also marked subtly different performances of health. In the Georgian period, drinking was the main mode of cure, whereas in the Victorian/Edwardian period the popularity of hydrotherapy meant an increased focus on bathing. Madden (1891) also records evidence of the transportation of Lucan water for use in Dr. Stevens Hospital in the city. Socially, Lucan mirrored wider cultural inhabitations of the watering-place. The original spa hotel had a large Ballroom which also served as a site of assembly while the walkways and grounds of both the Crescent and Lucan House served as the wider spa landscape. In the later manifestation, the more privatized setting of the hotel reflected the public-private shift in other locations (Gesler, 2003). In addition Lucan was acknowledged as being located in a therapeutic landscape setting in the Liffey Valley with access to social and pleasurable activities such as hunting, fishing and in later years, golf. The commodification of the spa was clearly one which at different periods did reward the entrepreneurship of the Vesey family and later, groups of businessmen/investors, though each was ultimately affected by wider shifts in society, culture and politics.

These more liminal performances of health and healing were reflected in the culturally created social scenery (Thrift, 2008). From the late medieval period on, the key to a spa town's success was to first attract visitors and once there, to keep them for as long as possible (Brockliss, 1990; Lenček and Bosker, 1998). Just as the symbolic and reputational powers of healing were metaphorically attached to the mineral waters, so the wider reputation as a place where 'society' met, mingled and reproduced (figuratively and literally), were crucial to the making of the spa town (Mackaman, 1998; Gesler, 2003). The Irish spa needs therefore to be seen in the context of the country's social and political circumstances of its time, especially in the colonial eighteenth and nineteenth centuries. The ruling Anglo-Irish were the class who were the principal creators of spa society. Prior to the development of spas in Ireland, the wealthier classes would have visited Bath, Bristol and even Spa as part of the Grand Tour (Kelly, 2009). With the creation of a domestic spa culture at home, there was an opportunity to reproduce the British spa in miniature. Thus the generic ensembles of the spa in England, with their seasonal attendances, accommodations, and curative performances were replicated in social practices such as Masters of Ceremonies (at Ballyspellan and Castleconnell) and assemblies and breakfasts (Mallow, Spa, Templeogue and Ballynahinch). Extra-curative activities included balls and dancing at a number of the spas with Pue's Occurrences noting in 1738 that: 'there is a spacious Long Room at Mallow ... lately built to entertain the nobility and the gentry during the season of the Spa. There will be Balls, Ridottos, Music Meetings and all other diversions as are at Bath, Tunbridge, Scarborough etc'. (Myers, 1984, 11). Here discursive colonial associations with English spas were recognizable, as were Rutty's accounts of the medicinal benefits of the Irish spas, as simultaneous and contradictory invocations of inferiority and superiority, a not unknown combination for those with identity and self-esteem issues, as the Anglo-Irish undoubtedly had (Bowen, 1942).

Many of the cultural performances at the spa were imbued with that sense of class identity, a reflection of the notion that practices enact identities and tastes, á la Bourdieu's notion of habitus (Bourdieu, 1977; Nash, 2000). Thus the behaviours were shaped by social relations and the kinds of activities that might be 'expected' in a British, French or German spa. These activities took place within and outwith the often fluid boundaries of the therapeutic place. The internal activities included social spaces where both genders could mix and apart from balls and regular dancing events, card-playing and gambling were a feature at Wexford and Castleconnell (Carroll and Tuohy, 1999). But the surrounds of the spa additionally had a more vigorously performative meaning as the location for all manner of field sports. Running the full gamut of 'shootin', huntin' and fishin'', and even cock-fighting (at Templeogue), these pursuits in Ireland operated as a discursive short-hand for the Anglo-Irish elite (Bowen, 1942). Hunting was popular at Ballyspellan, Ballynahinch, Mallow and particularly at Lucan, where proximity to the hounds of Meath and Kildare was a major selling point in its second phase of development (Madden, 1891). Fishing was popular at Spa, Lisdoonvarna and especially at Castleconnell, to the extent where the ultimate commercial future of that town was piscatorial rather than therapeutic (Dwyer, 1998; Hannon, 1984). Horse-racing was also popular, from the formal race-course at Mallow to the more informal Point-to-Point racing on the beach at Spa. All of these activities were a replication of society away from the spa, and so functioned as a reinforcement of identity for the Anglo-Irish, echoed in later periods by the appearance of croquet and golf at Ballynahinch and Lucan. 'Native' sports were rarely sighted though hurling is mentioned, appropriately played in front of the spa at Ballyspellan (Kelly, 2009).

Performances of health at the Irish spa were, as elsewhere, a mix of treatment and social spectacle. At the spa, mobility and rest were interspersed in the small daily rhythms of personal and communal movements. For every morning walk to the pump room, there were to be periods of rest and exercise, while many medical experts recommended a policy of 'early to bed, early to rise'. This, like many a routine and recommendation, was more honoured in the breach than in the observance. At Ballynahinch and Templeogue, the first waters of the day were drunk at eight in the morning, and elsewhere it was certainly drunk before breakfast (Handcock, 1899). The rest of the morning was given over to a leisurely and often communal breakfast, especially at Mallow, followed by walks and exercise. The water was drunk again in the afternoon, between one and three at Ballynahinch or from midday to five in Mallow. At Swanlinbar there was a specific routine observed in 1786:

> You go to bed at ten without supper. In the morning you appear at the well at six, drink till nine, taking constant exercise, and breakfast a little after ten. At one you return to the well and drink two or three glasses, returning home at three to be dressed for dinner at four. There is no particular regimen necessary but to be temperate in wine and to drink as little Chinese tea as possible. (Wilson, 1786 cited in O'Connell, 1963, 269)

At Lisdoonvarna, the need to mix the consumption of the water and the exercise of the body meant a spreading out of movement/rest in the course of the day: 'the proper dose of the Gowlaun water is from two to eight tumblers, or half pints daily, and the more divided this quantity the better, a stroll for some minutes intervening' (Horgan, 2002: 150).

While Swift's critique of the direct healing benefits at Ballyspellan have been noted, further critiques were to be found around the relative positions of the social and the curative. Many of these were recorded in traveller's accounts, as well as the more imaginative geographies of novel, poetry and song, all of which provided valuable insights into the social life of spas. They also functioned as good examples of the contestation of health and its relative position as performative act. Just as Cobbett famously satirized the gluttons and drunkards at Tunbridge Wells, so writers such as William Carleton and Elizabeth Bowen and travellers from Pococke to De Bovet, commented on the social behaviours at Ballyspellan, Mallow, Swanlinbar and Lisdoonvarna (Carleton, 1860; De Bovet, 1891; Bowen, 1942; Cobbett, 1967; Pococke, 1995). Excessive drinking, of both medicinal waters and wine, featured in several accounts from Castleconnell and Swanlinbar, reflecting contested narratives of health. In the case of the former, the 'seedy' feeling of the morning after was as often blamed on the water as the wine, while the powers of recovery associated with the waters were noted thus in Swanlinbar:

> A gentleman came to Swadlingbar with a set of company, with an intention to drink Claret more than Water and accordingly spent whole nights there in this criminal indulgence, in which he was too readily encouraged by the effects of Swadlingbar water as an antidote; for he affirmed that two or three quarts of it

taken next morning quite sober'd him, and made him as fresh and active as if he
had taken his natural rest; he also had a good appetite. (Rutty, 1757, 378)

Yet not all of the spas were wild places and Ballynahinch was specifically identified
as a sober location, in line with its Ulster Protestant and temperance links and a
reflection of local cultures of moral performance (McCullough, 1968).

The social reproduction of class identity was mirrored by its sexual equivalent,
and the function of the spa town as a setting for the pursuit of romance and
matrimony had a visible history as evident in Ireland as elsewhere (Austen, 1993;
Towner, 1996). In a country where the Anglo-Irish were a distinct though powerful
minority, the perpetuation of its elite class position was heavily dependant on inter-
marriage. The spa town was an important place in this process, and in an oblique
way had an additional performative and affective role as site of reproductive
health, with a literal romantic buzz in the air (Mackaman, 1998). The morning
assemblies at the earlier spas at Wexford and Mallow regularly featured eligible
army officers and the daughters of the landed gentry. The 'cures' at Templeogue
were alleged to work best on the 'young and healthy', while the infamous 'Rakes
of Mallow' were unfavourably described by a female visitor to the town in 1740s
who was affronted by their indecent talk. Carleton recounts in passing both the
curative, but predominantly social and matrimonial intentions of both male and
female visitors to the spa at Ballyspellan:

> There were also a great variety of others, among whom were several widows
> whose healthy complexions were anything but a justification for their presence
> there, especially in the character of invalids ... There was also a Miss Rosebud,
> accompanied by her mother, a blooming widow, who had married old Rosebud,
> a wealthy bachelor, when he was near sixty. The mother's complaint was also the
> spleen, or vapors; indeed, to tell the truth, she was moved by an unconquerable
> and heroic determination to replace poor old Rosebud by a second husband ...
> Wealthy farmers, professional men, among whom, however, we cannot omit
> Counsellor Puzzlewell, who, by the way, had one eye upon Miss Rosebud and
> another upon the comely-widow herself. (Carleton, 1860, 435-6)

The reference to healthy complexions, suggests early associations with beauty at
the watering place, where wellness and marriageability were equally served. The
reputation of Lisdoonvarna in particular as a 'matrimonial' site, extended into post-
independence Ireland and beyond (Furlong, 2006). As wealthy local farmers filled
the gap left by the Anglo-Irish after 1922, they sustained the tradition of going to
Lisdoonvarna at the end of the summer to find a wife, and this tradition has been
updated and commodified in a Matrimonial Festival held every September which
is a major feature of the town's contemporary identity and economy (Furlong,
2006; Keena, 2007).

Ultimately, cultural performances of health were central tools in the
construction and maintenance of the Irish spa. In this they enacted two main

functions, the spatial reproduction of English spas, and the social reproduction of the Anglo-Irish elite. In literal personal enactments of 'taking the waters' and in communal performances of class and stratification, the health and social meanings of the spa were inextricably linked. These were enacted in public, at the pump, ball or assembly, where the subtle behavioural distinctions of class elites could be observed. The aspirational enactments were also intended to imitate behaviours at Bath or Tunbridge Wells, sometimes even Baden-Baden, but had a secondary function in cementing Anglo-Irish identity (Kelly, 2009). In knowing the rules at the spa, either in terms of the number of glasses of water to drink or the recommended aspects of exercise and diet or indeed a lay knowledge of one's own condition and its monitoring, one's social position was demonstrated. In the wider society of the spa, the place of health also had a separate function as social glue, with which to bind that position into the future, through a 'healthy' marriage, which would further reproduce the dominant group (Duffy, 2007). At the same time and to the external eye, the spas continued to exist mostly as social centres and the playgrounds of the elites, where liminal behaviours and performativities of bodily excess, had a contested relationship with wellness and emphasized the conflicting relationships of healing and pleasure (Hembry, 1990; Durie, 2003b).

Spatial Performances of Health: Trams and Terraces

As in Britain, France and Germany, the spaces of performance were also significant in shaping health at the Irish spa. While Kelly (2009) noted the importance of fashionability, the ways in which the towns were constructed, accessed and utilized reflected additional spatial and relational aspects. As in Britain and France, the major spas were distant from the metropole but this physical distance was also discursively associated with a concomitant social distance (Towner, 1996; Mackaman, 1998). It was only the wealthy elites who had the money, and equally importantly, the time, to go to the spa for the season (Brockliss, 1990; Porter, 1990). Following the model of London, some of the earlier spas in Ireland were found in or around Dublin, at Chapelizod, Francis St, Goldenbridge and Templeogue. Nonetheless, the more distant spas at Wexford, Ballyspellan and Swanlinbar were considered the most important in the first half of the seventeenth century. Location and distance were sometimes spatial metaphors for exclusivity and exclusion (Gesler, 1992). Here the greater your social status, the farther you went from the core for your care and cure, while the liminal and liberating effects of distance were also important. Mallow and Ballynahinch performed this function for Cork and Belfast respectively.

While there was a spa in Galway and another short-lived development at Kilroran, it was telling that the West, traditionally the most Gaelicized parts of the country, was least penetrated by this cultural model of colonial urban development (Nolan and Nolan, 1992). Later spas were found closer to the cities as in the case of Castleconnell (for Limerick) and Lucan. Part of the reason why Lucan was popular, in both phases of its development, was its proximity to Dublin. It was far enough, and in terms of landscaping, rural enough, to meet those hydrotherapeutic requirements; yet it was close enough for a day trip. While this ran the risk of attracting a lesser class of clientele, the choice and quality of alternatives, unlike in the UK, were considered limited. Indeed many of the Dublin elite continued to frequent the spas of Southern England, especially Bath and Bristol, in considerable numbers (Kelly, 2009).

As was the case in most eighteenth century countries, the quality and spatial reach of the road networks were at best primitive. Turnpike roads were not developed properly until the early-nineteenth century while the canal network in Ireland had nothing like the same spatial extent as the British (O'Connor, 1999). Even in Britain the canal was not central to the operation of the spa town, apart from connecting Bath and Bristol (Borsay, 2000). Yet these limitations of transport did not deter the intrepid or wealthy traveller for whom this return journey was a small relative inconvenience of movement, compared to the extended period of rest which it initialized and completed. In the later period, the development of regular coaching services by Bianconi helped expand the reach of the spa while the role of the horse was important in connecting healing places and their users (Towner, 1996; Duffy, 2007). While the arrival of the railway in Britain was noted as impacting upon the spa town, in bringing in lower class clienteles, the arrival of the railway close to Lisdoonvarna was helpful; while the train and tram connecting Lucan to Dublin greatly enhanced its popularity in the 1890s (Joyce, 1901).

Spatial performances at the spa town were as much a question of who enacted them as much as what they looked like. Here the inhabited therapeutic landscape was a reflection not just of colonial elites but also of age, gender and health condition. The invalid and valetudinarian/hypochondriac were equally found at the spa, one legitimized through diagnosis, the other satirized for their unclear symptomatic histories. Yet both were legitimate representations of embodied health in place, reflected in their feelings and experiences. These different strands of personalized health narratives: real, imagined, assumed and experienced, all informed the inhabitation and construction of place. Young and old were welcome at the spa, though the presence of children is little recorded, apart from being part of wider family groups. Geographically the spas had catchments which were both national and regional. As noted earlier, the popularity of Ballyspellan and Swanlinbar was in part formed by the lack of viable alternatives elsewhere (Kelly, 2009). Yet as newer spas developed closer to the bigger cities, the clienteles became more regional such that Mallow developed a provincial catchment and Castleconnell drew from its immediate surrounds, not surprising given it was walkable from Limerick city (Hannon, 1983).

A greater mixture of society was noted at Ballyspellan and also at Lucan, Swanlinbar and Castleconnell, especially in the later nineteenth and earlier twentieth centuries, as Irish society itself became more democratic and less spatially stratified. Regularly recorded visitors in the eighteenth and nineteenth century were army officers, especially at garrison towns such as Wexford and Mallow. Yet an emergent category of visitor, increasingly visible and explicitly documented at the later spa from 1860 on, was the priest, a process symbolic of an invasion of colonial and exclusionary spaces by more native Catholic inhabitations (De Bovet, 1891). A link to earlier therapeutic practice was the presence at Swanlinbar, recorded in the eighteenth century, of pilgrims who stopped off en route to Lough Derg. There, the notion of the 'health pilgrim' was doubly enacted in place, with the mineral water cure used as a secular preparation for its spiritual equivalent (O'Connell, 1963).

Box 3.3 Lisdoonvarna, County Clare

Though copper, magnesia and chalybeate wells were recorded from the middle of the eighteenth century; maps from 1841 show only a few houses in Lisdoonvarna. The 1850s saw a rapid development and by 1859 there were almost 60 houses in the village along with hotels and two new sites, the Twin Wells and the Gowlaun or Sulphur Springs (discovered in 1852). In its early years, the local landlord, Captain William Stacpoole, helped promote the development of the town, but in 1867 he attempted to restrict access to the wells by building a wall and locked gate. This was blown up by an outraged populace and contestation of the ownership of the wells continued until 1915 when a local committee took over full ownership. As the spa town grew, social aspects developed in tandem with the curative. Hotels were built throughout the nineteenth century with evocative names such as the Royal Spa, Hydro, Spa View and Imperial. The peak years of Lisdoonvarna were from the 1880s to the 1920s with access augmented by improved roads, a steamboat service from Galway and the building of the West Clare Railway in 1887 (Hembry, 1997). In this period a number of landscaped walks and parks were developed for public assembly along with new treatment rooms at the Gowlaun spa. The arrival of the then Lord Lieutenant of Ireland, to re-open the newly refurbished Thomond Hotel in 1906 marked a highpoint in the social status of the spa. After World War I, traditional utilizations of the spa by the Anglo-Irish elite declined, though the strong farmer and professional classes of the new state sustained the spa into the 1940s. The pump room is still open in the summer, but it is the social reputation of the town, in particular the Matrimonial Festival in September, which sustains the town's reputation (Furlong, 2006).

Figure 3.4 Panorama of Gowlaun Spa, Lisdoonvarna, Co. Clare, c1910
Source: Courtesy of the National Library of Ireland.

As the only town in Ireland with multiple wells, Lisdoonvarna developed a strong curative reputation linked to different place-cures and treatments. The copperas waters were, wisely, used externally for ulcers, sores and skin diseases. The original Double Wells (magnesia and chalybeate) were used as blood tonics and for weak stomachs. The unusual Twin Wells (with separate but close sources of sulphuric and chalybeate water) and the main sulphur well at Gowlaun were considered effective in curing gout, rheumatism, consumption and liver complaints. The different wells, spread out around the town, also gave rise to a mobility of utilization with 'health consumptions' interspersed with social interactions. Given the relatively late development of Lisdoonvarna, it adopted newly popular practices of hydrotherapy, aided by a supervisory spa doctor from the 1870s. New treatments included, 'the needle bath, the general electric; the Schnee bath (Galvanic, Faradic or combined), Ionization and electro massage. Various medicated baths as prescribed, are also given, such as Acid, Alkaline, Bran, Pine, Peat, Nauheim.' (Dooley-Shannon, 1998). Despite the presence of these newer medical agents of control, the town still had a liminal edge, of which the contestation of ownership of the water source and the wilder social performances played a part (Lynd, 1998).

The built environment was also developed to try and replicate British spaces in the therapeutically specific ancillary buildings and accommodations that spa towns were expected to have. A material shorthand of architectural design, bricks and mortar, reproduced the metaphors of place and health (Geores, 1998). These elements were at heart cultural, with attempts to recreate in built form a whole range of symbolic and colonial meanings (Gesler, 2003; Borsay, 2006). Each of the spas had some form of pump house from which the water was distributed. At Ballyspellan, the spring itself was around a mile and half east of the town on the upper slopes of Spa Hill. Here a now-ruined stone spa-house stood, with the spring itself still visible in a small stone enclosure which, given the presence of statues and rags, one could even assume to be a holy well. At Ballynahinch, one of the original hexagonal wooden buildings remains, surrounded by undergrowth, while the spa well at Castleconnell was covered by a classical stone enclosure with a copper roof in 1759. The spa-house beside the River Liffey at Lucan survived for almost two centuries before being swept away in a flood in 1954. Quite apart from the spa-wells themselves, the other features of the classic British spa included Assembly Rooms, treatment rooms and baths. It was the relative shortage of the latter, apart from the unique open-air baths at Leixlip, that was the most marked difference in Ireland, prior to the popularization of hydrotherapy. Meeting halls of variable size were also developed at Ballyspellan, Lisdoonvarna, Lucan and Swanlinbar. A small stone building, marked by distinctive acorn finials, enclosed the well at Spa, although its social activities were centred on the town of Tralee, three miles away, where there were, 'good shops and good society' (Rutty, 1757).

Outside of the 'spa-specific' elements, the spa towns were also settings where the landscaping was literal. Here the requirements of the '(sub-)thermal economy' meant a demand for ancillary accommodation, leisure and pleasure facilities (Mackaman, 1998). The architectural legacies of terraces, arcades, parks and walks were found in Ireland, if in miniature. Hotels were built specifically to house spa visitors at Swanlinbar, Lucan and Ballynahinch (a mile and a half from Spa village) while other towns such as Mallow and Wexford had plentiful accommodations. New terraces were built at Mallow, alongside the canalized water supply from the well, while a well-built linear block, named in a discursive twist as 'The Crescent', was built at the rear of the original Lucan Hotel in the 1790s. In the case of Ballyspellan, the original lack of accommodations and lodgings was solved by the simple expedient of building a new village, Johnstown, at the bottom of the hill on the main turnpike road from Dublin to Cork (O'Ferghaill, 1987). Even then the demand in season was such that every house in the village and surrounding countryside was full (Knox, 1845). In Lucan's second phase the relationship between spa and hotel was simplified by co-location. An elegant new pump room, hexagonal in shape, was built in 1891 with a direct piped link to the well at the river and formed one wing of the newly created Spa Hotel (see Figure 3.3). In this case, the built environment of the spa was a direct material consequence of the metaphor of place=health, neatly encapsulated in a single building (Geores, 1998). In a performative sense, many of the hotels at Lisdoonvarna had glass verandas along the main street, while at Mallow, a much mentioned feature (of which some remain), were the projecting windows on the first floor of the main street (Lewis, 1837; Dooley-Shannon, 1998). These were 'voyeur-flâneur' spaces from which one could see or be seen and were allegedly designed specifically so that young women could check out the available men as they promenaded to the spa and back (Myers, 1984).

There was a marked spatio-temporal performance in place as in other therapeutic landscapes. Each of the spa towns had a marked season, typically from May to October, with early starts in April noted at Mallow, Swanlinbar and Templeogue. Lucan, in its later nineteenth century reincarnation, had a year-round season, though summer remained the most popular visiting time. The end of harvest and rural rhythms of the west dictated that the matrimonial festival at Lisdoonvarna took place in September. The shifting role of time within the spas was also linked to the arrival of the railway (Mackaman, 1998). The railway partially opened up access to the spa towns at Mallow, Ballynahinch, Lucan and Lisdoonvarna, but the meaning intended by Mackaman was to draw attention to a shift in the ways in which time was understood and enacted. Here a mechanical clock acted as a metaphor for the replacement of natural temporal performances of health to more socially managed timetables of curative treatment and formal dining (Mackaman, 1998). However there is little evidence that such strictures of control were entirely successful in the Irish spa town, prior to the 1850s at least.

As a spatial model of therapeutic performance suggests, places were also subject to material and metaphorical effects from beyond their bounds and the Irish spa was equally susceptible. The setting of the Irish economy as part of, but subservient to the British, was reflected in a mobility of fashion from beyond the island's shores. It was no co-incidence that the heyday of the spa in Ireland was between 1780 and 1820, a period which coincided with a reduction of access to the Continent associated with the Revolutionary wars in France (Kelly, 2009). After this period and more intensively from the mid nineteenth century on, the fashion for sea-bathing and the revived popularity of the coast also caused a visible spatial shift from interior (where almost all of the spas were located) to coast, associated with these therapeutic performances (Urry, 2002; Durie, 2003a; Kelly, 2009). The role of Bray in affecting the business of Lucan or the impact of Newcastle and the coastal resorts of West Clare on Ballynahinch and Lisdoonvarna respectively, were examples of this wider global shift in leisure activities (Urry, 2002; Kelly, 2009).

Nature and culture were markers within a continuum of performative spaces. The affective power of nature and landscape was informed by broad suggestions within accounts of its restorative and restful links to healing/wellness (Tuan, 1974; Conradson, 2007). In the selling of the spa, either through the guides of Rutty or Knox, or from the traveller's accounts and local literatures features such as views, landscape and topography was explicitly used in the selling of place. The coastal location of Spa provided extensive views of the Dingle Peninsula and was praised as therapeutic; as were the rolling hills and trees around Mallow and Lucan and the beautiful riverside setting of Castleconnell. Knox noted the 'medical topography' around Mallow and explicitly describes the value of landscape at Ballynahinch, incorporating its immediate surrounds in a multi-scale and relational appropriation of space:

> The pump rooms are two neat octagonal buildings ... situated in the centre of a variety of pleasant walks, ornamented with planting, which afford to the visitors agreeable promenades, in the intervals of water drinking. ... As the character of this place has grown up gradually, so every thing points to the increase and permanence of its reputation. Its pure, dry, bracing mountain air, its elevated situation, quiet but cheerful retirement – its excellent tonic and alterative – the favourable opinion of the neighbouring physicians – the vicinity of Belfast – the rapidly growing metropolis of the north, will, I doubt not, long combine to render Ballinahinch, a favourable resort of the inhabitants of the north-east of Ireland. (Knox, 1845, 269)

In addition, and as a foretaste of its later utilization at the sea-bathing resort, the 'bracing air' at Ballynahinch and Ballyspellan began to be incorporated into therapeutic narratives of place.

A consideration of spatial performances of health has merit in its application to the spa town, especially when understood in relational terms (Cummins et al., 2007). Mobile populations, elite or otherwise, enacted mobile curative and social practices in place, while the popularities of individual towns were in turn affected by the unpredictable whims of fashion. But the spas were also mobile built environments, reflected in the morphological shaping of small spa 'quarters' in towns like Mallow (see Figure 3.1), Lucan and Wexford and the creation of new towns at Johnstown (Ballyspellan) and Lisdoonvarna. In these settings the material built environment was directly structured on the metaphoric associations with health (Gesler, 1998; Geores, 1998). While clienteles shifted backwards and forwards in line with the shifting identities, loyalties, exclusions and positionalities of the Anglo-Irish, there were increasingly frequent incursions and utilizations as accessibility broadened to the wider Irish population after 1922. While the geography of the Irish spa town was not comparably in terms of its scale to the larger spa cultures of England, Germany or France, it was in its reproductions of the processes of health and culture in those places (Kelly, 2009). In addition the role of place in shaping health experiences and performances also highlighted those inhabited dimensions of the watering-place. Just as naming could be norming, so too could moving and dwelling (Berg and Kearns, 1996; Wylie, 2007).

Economic Performances of Health: Commodification and Celebrity

Underpinning the complex interweaving of health, culture and place were wider economic performances of health (Dorn and Laws, 1994; Saks, 2005). The roles of entrepreneurs and developers were crucial to the successful development of spa towns (Geores, 1998; Borsay, 2000; Gesler, 2003). Yet often, those same entrepreneurs and developers were involved in less frequently reported stories of decline and failure (Towner, 1996). The narratives of spa town development were broadly typical, though there were of course unique variations. In Germany, Canada, and to a lesser extent, France, there was a more evident 'state' involvement in the construction and development of spa towns (Wightman and Wall, 1985; Hahn and Schönfels, 1986; Brockliss, 1990). Elsewhere free market capital and private investment were the main structural factors in the making of place, as in Britain and Ireland. Typically, the same cycles of discovery-test-verification-development were encountered, though there is evidence from Britain of a more retrospective order of construction of resort and waters (Hembry, 1990). Nonetheless, the Irish spa seems to have followed the normal route. So at Mallow, Castleconnell, Ballynahinch and Ballyspellan, the publishing of chemical test results on the mineral contents of the waters by the likes of Rutty, was a key first step in the production of a healing place. Doctors provided the medical evidence needed to verify and authenticate the spa's effectiveness through patient biographies at Spa, Swanlinbar and Lucan (Rutty, 1757). In time this hardened into a link between medical experts, local landowners and additional commercial interests in the form of hotel owners, trades people and local investors, especially in the earlier spas at Mallow and Swanlinbar. Finally as the role of the landlord declined, private business, more powerful and culturally embedded medical interests and even local government formed the key alliances in later spas at Lucan, Ballynahinch and Lisdoonvarna (Madden, 1891; Furlong, 2006).

As initial forays into health tourism and place promotion, the business of health was central and;

> The pattern and pace of development among watering-places depended not
> only upon their ease (or otherwise) of access, but also upon how they catered
> and competed for the patronage of the visitor population. It appears that resorts
> originated in places where economies were particularly open to change, and
> that they developed a culture especially conducive to entrepreneurial activity.
> (Borsay, 2000, 783)

While place and clienteles mattered, it was the mobile entrepreneurial cultures
that made the difference. There were variations in the 'entrepreneurial therapeutic
alliances' in different locations as the development of a spa as a going concern
required risk, capital and investment. In many places that risk paid off in the short
term but was not sustainable. Thus the spa at Wexford was short-lived but was
developed in pre-capitalist times around 1700. Mallow's foundation featured an
alliance between local landowners, the Jephson's, and a medical man, Dr. Rogers
(see Box 3.1). At Ballyspellan, several doctors worked with John Hely, the local
landowner who invested in very practical terms by constructing a village near
the spa in the 1770s. Here that alliance was a quite literal example of health/
place production. The investment was not an enormous success and by the 1870s
a local committee ran the spa (Kelly, 2009). The history of Lisdoonvarna was
also marked by shifting mobilities from private/individual to public/communal
investment as the original owners and developers, the Stacpoole's gradually
gave way to Local Improvements Committees (1880) and Associations (1915)
(Dooley-Shannon, 1998).

The wider commodification of the waters also meant that there were different
ways of making money, causing agency to shift beyond the health-specific spa
facilities to the provision of ancillary services and entertainments. While focused
on the waters, recurrent tourism approaches were interested in getting visitors
to the 'attraction' and keeping them there (Urry, 2002; Smith, 2003). In season,
visitors created a demand for accommodation, food and transport. In the early
smaller spa resorts this meant that every house was put to use as accommodation
and lodgings (with food) as noted at Ballyspellan and Swanlinbar (O'Connell,
1963; O'Fearghaill, 1987). At Spa and Castleconnell, a number of large villas
were built, while the terraces at Lucan and Mallow, constructed by Vesey and
Jephson, the respective local landowners, were driven by simple contingencies of
supply and demand. A Mr. Gabbett constructed a block of elegant houses (later the
Tontine Inn) at Castleconnell as a speculative and failed venture in 1812 (Herbert,
1948). Finally, there was a form of 'trickle-down' effect to the normally excluded
native populations of the spa town in the form of ancillary employment, much
of it carried out by women, such as domestic, food, trades and transport work
(Dooley-Shannon, 1998; Borsay 2000). In the case of the latter, argumentative and
rapacious 'carman' were recorded at Lisdoonvarna and Ballynahinch where they
ferried visitors from respective steamboat and railway terminals (Horgan, 2002).

As noted at the holy well, the water was both a potable and portable product.
But product it certainly was and as such was central to the commodification of

health performance in place. It also made more explicit modalities of production and consumption. Here the product, whether the sulphurous draught of Swanlinbar or the light chalybeate of Ballynahinch, was in the form of the water itself and visitors were specifically encouraged to drink regularly and over an extended stay at the source. There was a tension associated with this product/place dichotomy, around whether the product should be consumed in place as opposed to taken away. Indeed Rutty's encyclopaedic listing of the mineral waters of Ireland was explicitly classified according to whether the water could 'bear carriage' or not. So for example the sulphurous waters of Swanlinbar stood up well to transport whereas the chalybeates of 'the first class', which included Lisdoonvarna, could only be drunk at the source (Rutty, 1757). However for the entrepreneurial development of the spas, this was a problem. If the water could be bottled and moved then this obviated the need to visit the source (Brockliss, 1990). While money could be made from the sale of bottled water, much more money could be made at the spa. As a result, considerable effort went in to convincing users of the need to consume the product in the place for maximum benefit. Despite these attempts to discursively tie water to place, there were early references to the business of selling bottled water. Kelly noted the presence of (Belgian) Spa water in Ireland from as early as 1700, while from 1732 at Templeogue, there was a steady supply to Dublin markets with a number of city centre suppliers including John Brown's tobacconist in Crane Lane; Mallow water was also sold in Cork in 1758 (Kelly, 2009). Significantly, the product was specifically associated with its curative outcome; '... the curative powers of water have been widely celebrated, often engendering ferocious local disputes as to the desirable mineral constituents of particular healing springs, wells, streams and spas, and the precise technologies of expert treatment appropriate for diverse diseases and particular cases.' (Porter 1990, viii).

Here the processes of treatment and cure were as noted, contested by place as well as increasingly by the forms of treatment, but the mention of ferocious disputes around the waters confirm the importance of competition, ownership and marketing at the spa.

A modern notion of 'celebrity endorsement' was also extensively used as part of a linkage between fashion, power and patronage. As in Britain or France, where visit by members of royal families boosted the reputation of healing places such as Leamington or Plombiéres, so in Ireland aristocratic and power connections were important (Denbeigh, 1981; Brockliss, 1990). As the colonial de facto 'head of state', the Lord Lieutenant's visit to Lisdoonvarna in 1906 gave it a new lease of life to that and subsequent seasons (Furlong, 2006). There was also an amount of snobbery associated with the social cachet of a spa, and the quality of its visitors was often discursively elided with its quality as place. Later accounts of spas such as Swanlinbar, Mallow and Castleconnell, almost invariably representative of the 'Orientalist' colonial view and governed by assumptions of the 'virtue' of class, all commented on the presence of the 'lower ... and lesser' classes of society as a sure indicator of decline (Hannon, 1984; O'Connell, 1963).

The medicalization of the product was also significant. In framing these as performances of consumption, the reach of the medical profession was extended to the wider spa treatments. While these were associated more with the later spas, the role of the spa doctor, especially at Lisdoonvarna and Lucan, was important in widening the commodification of health in place. While never as powerful as the *intendant* of the state-run French spa, the spa doctors at Lisdoonvarna had their own large home and surgery/clinic, 'Maiville', directly above the spa (Mackaman, 1998). Through a combination of consultations, referrals and new hydrotherapy treatments, they exerted considerable power and derived profit from the spa in the later nineteenth century (Dooley-Shannon, 1998). As the effective gatekeeper of access, they marked a subtle shift of power and ownership in the healing business of the spa. This was a process created discursively outside the spa through the professionalization of the practice and more importantly the reputation of scientific medicine. In this the assumptions that the sustenance of the spa town was down to social fashion alone must be set against these underlying structural forces of the medical gaze and its transformation of health from aspirational trade and patient/ doctor negotiation to a capitalist business exercising economic and embodied control over subjects (Foucault, 1976).

As with any commercial product, the variety of business interests involved invariably gave rise to contestations of ownership, expressed spatially in the product. Given their commodification as 'healing waters', the ownership of that water was a significant factor in the production of place. As such it needed to carefully balance opposing tendencies of access and exclusion, public discursive ownership and private appropriation and the opposing forces of greed/profit against a public right/good. This tension between appropriation and public rights of access was exemplified in Ireland by the failed attempts of the Stacpoole family to restrict access in Lisdoonvarna in 1867 and 1897. Captain Stacpoole enclosed the source and placed locked gates at the Gowlaun pump in 1867. In this case, the locals simply blew up the walls and gate, while the later dispute went to the courts and eventually saw the transfer of ownership and management of the spa to a local committee (Dooley-Shannon, 1998). Neither event could be entirely separated from underlying political sentiment and in this case the contest over health access was symbolic of the wider anti-colonial struggles of the Fenian and Land League eras (Dooley-Shannon, 1998; Dwyer, 1998). Vernacular access to the spa waters immediately outside the spa 'domain' was provided at the Dog's Heads in Mallow in 1857, which may be interpreted as simultaneously opening up access while keeping local users out (Jephson, 1964).

In considering the role of ownership and power from a patient perspective, these wider regulations and medicalizations of treatments, marked an important shift in more experiential and affective self-place health encounters (Porter, 1999; Conradson, 2007). The power relationships between patient and doctor were being shifted from the mid nineteenth century on. Despite the identity of hydrotherapy as a form of CAM, it was, in its treatments, practices and imagery of machines and white coats, indicative of a more regulatory and managed form of care; in which the patients had less and less of a direct and active role. As the spa doctor assumed a greater importance at Lisdoonvarna, the older and more autonomous practices of drinking, walking and bathing became negotiations of health between consumer and producer. At Lisdoonvarna, a range of new curative treatments and hydrotherapy baths emerged (see Box 3.3). In these mixtures of exoticized and vernacular treatments, one can see a new form of mystery cure emerging. Instead of the magic cure of the well, the new magical narrative of medicine are used to shape patients practices, and in so doing defer their self-managed and owned health into the hands of expert others (Porter, 1992).

The decline of the spa town in Ireland was due to a mix of spatial, cultural and embodied reasons, but the most significant, according to Kelly, was, 'the credibility of the water, which was elusive and essentially unpredictable' (Kelly, 2009). As the fashions of society were aligned and co-dependant on the fashions of health, there was a mutual connection between health and pleasure in the service of well-being (Hamlin, 1990). These 'therapeutics of place' were part of both a metaphorical telling/selling and a material construction of place (Cantor, 1990). The spa towns developed a discursive commercial meaning with aspects of marketing, advertising and the consumption of product (in the form of bottled waters) acting as what Geores noted as an early example of commodity fetishism (Geores, 1998). While entrepreneurship was essential to the development of the spa town, it was on its own, no guarantee of success. Conversely many of the towns that had inherent natural advantages were left down by a lack of commercial nous (Borsay, 2000). These economic aspects of the production of the therapeutic places were, when combined with the more inhabited notion of consumer confidence, indicative of a certain 'spatial chemistry' which often determined their success or longetivity. It was certainly suggested that in better hands than the Jephson family, Mallow might have prospered more than it did; while the natural advantages of Leixlip over Lucan were undermined by the more developed infrastructure of the latter as well as its greater proximity to Dublin (Madden, 1891). Finally, as with economies everywhere, the development of new fashions for healing waters, especially sea-bathing, were crucial factors in the decline of the spa (Borsay, 2000; Durie, 2003a). From an individual health perspective, the developing business of health and its regulation and management in place, altered previously freer ownerships and performances of health and self-treatment.

Summary: Cure and Communitas

By circulating reports of the numerous cures effected; and by advertising the delights of the town and neighbourhood as a health resort for invalids, they sought to rival the famous English city in the number of visitors attracted thither. After a time, promenades were constructed, new walks planned and bands discoursed music, while the company drank at the wells. Assembly rooms for card parties and dances were opened, and during the season, life would seem

to have gone on as it has always done at resorts of the kind – only no Beau
Nash arose to make Mallow the fashion, like its English sister, and another Jane
Austen was needed to lay at the Spa the scenes of stories inimitable as those she
wove around Bath. (Berry, 1892, 111)

In developing the spa town in Ireland, different performances of health were
riven with the ambiguous identity of the Anglo-Irish class. In the narratives of
health at the spa town, other ambiguities were evident. In purely medical terms
these were reflected in chemical and medical arguments over the content of the
waters and their curative outcomes and meanings (Coley, 1990). Rutty's lengthy
documentation of these elements pointed to their importance in the sustenance
of spa town reputations (Rutty, 1757). More broadly, the curative and therapeutic
reputations of the spa town were also measured against their social attractions.
The contested health and anti-health performances and practices, from the Rakes
of Mallow to the match-makers of Lisdoonvarna, mirrored wider debates which
saw in the pharmaconial nature of the waters and associated performances, an
important shift in how inhabitations in place were as important as the places
themselves (Conradson, 2005a; Collins and Kearns, 2007). They also provided
a valuable therapeutic link between the holy well and later hydrotherapeutic
practices in their consideration of the moral geographies of the body. Here
embodied ownerships of one's own health were becoming regulated from
outside, as the more medicalized developments at Lisdoonvarna and Lucan
impacted and to an extent, invaded that personal body space. Finally, one
could see in the contestations of the curative and the social, further metaphoric
connections to the liminal and the carnivalesque (Bakhtin, 1984). While the
'water pill' continued to be the unpleasant draught of Swanlinbar, it was also
in the metaphor of candied orange peel, sugared for public consumption in a
range of discursive forms. In seeing the spa town as a 'sugared place' in moral
geography terms, one could also see the critiques of medical performances and
behaviours to come, especially at Ballynahinch where a strong Protestant ethic
of abstemious behaviour contrasted with other towns (Mackaman, 1998).

The spa towns were culturally constructed entities which, though individually
conceivable as ephemeral, remain a collective part of the Irish landscape for over
300 years. While Lisdoonvarna is the only town where one can still formally take
the waters, there are relict material landscapes of terraces, springs, spa-houses,
wells, walks and hotels. The discursive connection between building and town also
lingers in names. Wexford still has a Spawell Road while two separate villages at
either end of the country bear the name Spa as does the hotel in Lucan. While the
curative functions of the mineral waters are less credible in contemporary formal
medical spaces, the waters can still be accessed and drunk in place at the top
of Spa Hill outside Johnstown or at Dromod Sulphur Spring east of Swanlinbar.
Other spa-elements are now to be found within private residences as is the case of
the spa well and assembly rooms at Templeogue and Castleconnell respectively.
While the material business of illness and its treatment are gone, as places the spa

towns sustain echoes and traces of their material and even affective pasts (Joseph, Kearns and Moon, 2009).

As embodied spaces of health performance, the Irish spas also mirror the cultural production of healthy places in other jurisdictions. Formal medicine regularly contested the 'authenticity' of the kinds of health provided in the spa town, in particular through accusations of quackery and the liminal natures of cure, curer and even curee. It could be argued that such points of view were justified given the ultimate decline of the spa. Yet the discursive and material value of the healing waters of the spa retains its strength in countries such as Germany, Hungary, Italy, Japan and Korea, where traditional bathing in and consumption of mineral waters remains an everyday occurrence (Smith and Puczko, 2009). It should also be noted that the forms of treatment and the current medical understandings of their time, placed spa towns at the centre of early forms of health tourism. For much of their history, the health, and as importantly, the reputation of health was always as important as the social attractions (Towner, 1996).

As a commodity, there were strong associations with emergent scientific constructions and commodifications of health that differed from earlier faith narratives. The discovery of lithium at Lisdoonvarna in 1875 even flags up a strong link to a treatment still widely used in contemporary mental health care. The chemical analyses and narratives of cure created discursive links between science and the healing power of place. These were enthusiastically marketed and promoted by a wide range of actors, some with more structural power than others, but all engaged in a wider mission to sell the Irish spa to its own people. It had more limited aspirations than other colonies such as New Zealand, where that mission was part of a national tourism initiative; ironically such a vision was later associated with Irish sea-bathing places (Rockel, 1985; Furlong, 2009). The realization and success of these commercial ventures were at one glance ephemeral, but in maintaining a popular spa for 20 or 30 years, the entrepreneurial alliances at supposedly failed spas such as Swanlinbar or Wexford, matched the life cycle of equivalent local spas in Britain and the Continent (Palmer, 1990; Hembry, 1997). The wider failure was in the flawed vision of matching or competing with larger national models such as England or Germany. In a consideration of therapeutic landscapes, the acknowledgment of metaphors of the relative over the absolute, of the relational over the scalar, may help in better situating the Irish spa town (Cummins et al., 2007).

In their clienteles, developers, servants and commentators, the spa towns also functioned as inhabited and experiential landscapes of health and healing (Wylie, 2007; Lea, 2008). The material/metaphoric productions of health were exercized through embodied acts of ingestion and to a much lesser extent than elsewhere, immersion. The place of the body also shifted across time from a freer performativity to subjection to a more medicalized gaze, though this latter form was less fully developed than in the later hydropathic period (Mackaman, 1998). The commodification of the spa town also had a visibly inhabited dimension. In season the spa town cultures of product, treatments, transports, lodgings and entertainment sprung into life with a mobile and inhabited script of class, religion and politics. Ownership also featured in the contested cycles of political and commercial power and in access to and utilization of the waters of the spa. As well as mobile inhabitants, there was also a discursive mobility – of ideas, accounts and stories – reinforcing place (Towner, 1996). Even in 2009, older residents take away the spa water of Lisdoonvarna for domestic use, sworn by verbally as an effective treatment for backache and rheumatism. The role of the spa town as liminal place and even as asylum, in the form of escape from the everyday, was also bound up in these seemingly contradictory conflations of health and pleasure.

As noted by Conradson, one cannot assume that healing and restoration were only achieved in silence and for some the value of the social, in the communitas of the spa, was as valuable as the physical cure (Turner, 1973; Conradson, 2005b). Here the performances of the spa society, so long condemned in satirical imaginative accounts for a lack of healing values, can be reassessed in more contemporary terms for their value as sites of sociability and collective benefit. Yet in Lisdoonvarna, the social importance of a good 'yarn' drew crowds of rapt listeners and was considered to interfere with the performances of taking the waters (Dooley-Shannon, 1998). At the spa individual performances of health were shaped by their collective identities which reinforced and validated that enactment. Such performances were positioned within a more subtly constrained production of health in place and reflected a; 'much more anonymous and as it were horizontally distributed exercise of power; power exercized both over oneself and between and across selves ... a changeable matter of elaborated cultural conventions and ramifying codes of conduct' (Matless, 2000, cited by Wylie, 2007, 112).

Here the power one had over one's own health was shaped by conventions and communal beliefs, which were suggested as being unconstrained by rule and regulation. Yet they bore the bonds of restraint in personal psycho-cultural ownerships and expressions of health. Finally, new forms of medicine associated with the developing popularity of both sea-bathing and hydrotherapy were destined to have an effect on the healing reputation of the spa (Urry, 2002; Bergoldt, 2006). At times the relationships between these three different therapeutic landscapes were explained as a successive process; the next two chapters however, examine in more detail the complex cultural constructions of those forms, as well as material diversities of introduced and vernacular hydrotherapeutic landscapes.

Chapter 4
Turkish Baths and Sweat-houses: The Sweating Cure

Introduction: Heat and Cold

> I had heard of a peculiar practice of the inhabitants of this part of the country ...
> I refer to what are called 'Sweating Houses' ... it is a species of oven five or six
> feet high by about three in width, with a hole for entrance of about one and a half
> feet high at the level of the earth, the whole construction being the shape of a
> thimble. To use the sweating-house they heat it with turf, exactly in the way such
> a construction would be heated for the purpose of baking bread. When it is pretty
> hot, four or five men or women, entirely naked, creep in as best they can through
> the little opening, which is immediately closed with a piece of wood covered
> over with dung. The unfortunates stay in this for four or five hours without the
> possibility of getting out, and if one of them takes ill, he or she may sit down, but
> the plank will not be taken away before the proper time. As soon as the patients
> enter, an abundant perspiration starts, and commonly, when they come out they
> are much thinner than when they went in. Wherever there are four or five cabins
> near each other there is sure to be a sweating house, and no matter what may be
> the malady of the peasant, he uses this as a means of cure. (De Latocnaye, 1797,
> cited in Weir, 1989, 11)

Tramping across remote fields in the border counties of Cavan or Leitrim, one
occasionally comes across what looks like a small stone igloo, usually with a
covering of sods and generally in a ruinous condition. These sweat-houses are
a relict reminder of a rural sauna; an Irish version of a global sweat-cure. One
might also see a small stream or plunge pool alongside suggesting that water
was centrally involved in this performance of health, being drawn from the body
through sweat and providing re-hydration in the subsequent plunge. In these core
elements of perspiration and bathing, connections can also be traced to warmer
enactments within Roman baths and Arab cultural traditions associated with the
hammam or Turkish bath (Croutier, 1992).

   The well and the spa were both related to natural sources of water. This
chapter looks at two less natural but linked forms of 'healing waters' found in
Ireland, namely, Turkish baths and sweat-houses. As primarily indoor settings,
they represent more intimate and private spaces than the public settings of
the holy well or spa. Both places had strong associations with water, albeit in
different forms than previously encountered (Shifrin, 2009). They reflected

the ephemeral flow of water itself in time and space, but also in and out of the human body in reflexive acts of permeability and porousness (Smith, 2005; Strang, 2008). This porousness was integral to the performed water cure, where the emphasis was on sweat, detoxification, hydrotherapeutic treatments and the 'hot-air bath'. Such cures were originally linked to Greek hypocausts, consisting of dry heat and cold water plunges; while the term *laconium*, as a proto-sauna, was later used in Roman baths (Buckley, 1913; Jackson, 1990). Other places where sweating and detoxification provided a healing function were to be found independently in the saunas and *banias* of Northern Europe and in the sweat-lodges of Native American and First Nation tribes in North America (Aaland, 1978; Geores, 1998). In the case of the latter, sweat-lodges had additional spiritual and symbolic meanings associated with cultural performance and identity (Wilson, 2003; Weir, 2009). Other global examples include the *temazcalli* (translated from the Nahuatl for 'bath' and 'house') of Mexico and *hammams* (Guerra, 1966; Aaland, 1978). The Turkish bath had a particular association with Dr. Richard Barter of Blarney, who designed and launched an innovative and much-copied design in 1859. This later developed into a functional franchise for his patented 'Improved Turkish bath' across Ireland, Britain and even Germany (Murphy, 1979). In contrast, sweat-houses were small, vernacular saunas used to cure rheumatic and fever-related illnesses in upland areas (Evans, 1957; Weir, 1979).

In considering the 'sweat-cure', there was a direct healing link from the spa town through the Turkish bath to the development of hydrotherapy from the middle of the nineteenth century (Price, 1981; Whorton, 2002; Durie, 2003a). Traditional embodied performances of health such as drinking and bathing were augmented by internally produced water-cures in perspirative and invasive forms and a variety of new technical treatments (Mackaman, 1998; Bergoldt, 2006). The rise of hydrotherapy acted in a schizophrenic way as both a resistance to, and a support for, the tendency toward greater medicalization in the period after 1830 (Porter, 1990; Wear, 1992). As that medicalization was often characterized in the pharmaceutical development of the 'magic pill or bullet', many therapies of this period, of which hydrotherapies were the most prominent, were more natural responses (Porter, 1999; Borsay, 2000). In countries such as Germany and Austria, early proponents of hydrotherapy like Priessnitz and Kneipp, promoted the cold-water cure as a wider complementary and holistic medicine, where dietetics, exercise and hygiene played equally important roles (Price, 1981; Bergoldt, 2008). These lay understandings of self-care were by their nature a challenge to the growing power and prestige of formal medical bodies. Priessnitz, Kneipp and Barter were successful entrepreneurs and their presence as real economic rivals to the rapidly-establishing medical professions lay behind attempts to contest their therapies as dangerous and irrational (Heller, 2005; Bergoldt, 2008). However, the often sadistic and harsh technical treatments used in hydrotherapy were simultaneously associated with an increasingly authoritarian medicalized gaze (Foucault, 1976; Mackaman, 1998).

Sweat-houses were generally referred to in Irish as *Tighe Allais* or *Alluis*, a direct translation of 'sweating houses' (Weir, 1979). No longer used since the end of the nineteenth century, their origins and longetivity were the source of some debate and little agreement (Buckley, 1913; Evans, 1957; Weir, 2009). They had a specifically restorative/healing function and were primarily found along the border between the Republic and Northern Ireland (Logan, 1981). While they were often referred to as 'hot-air' baths, their operation was broadly similar to a Finnish or Swedish outdoor sauna (Aaland, 1978). Carbon dating evidence suggested they date from the fifteenth century, but their fullest development was in the late eighteenth and early nineteenth centuries. They were an unusual example of an indigenous healing place and while it is rare for 'folk' medicine to be given the same valorization as indigenous people's (often discursively read as non-western, non-white) health narratives, one could critically question this positionality, especially given reference in colonial texts to the Irish as 'aboriginals' (Curtis, 1997; Kavita, 2002). However folk-medicine's strong associations with land, water and nature, do place the sweat-house squarely within wider indigenous therapeutic landscapes research (Wightman and Wall, 1985; Geores, 1998; Wilson, 2003).

Theoretical concerns with performances of health are especially relevant in these 'sweating-places' (Lorimer, 2005; Lea, 2008). As a cure, which in essence emerges or is flushed out of the body, the physicality of that experience can be both oppressive and invasive, but is always individually experienced. In biomedical terms, the vital function of the skin as excretory organ is well-established (O'Leary, 2000). In its sensual exposure to settings, which have spatial physicalities of heat, dryness, steam and humidity, it is reflective of water's own transformative qualities (Rattue, 1995; Strang, 2004). The settings were also affective places, where mood, lighting and the sometimes exotic surrounds and design elements all combined to take the patient (and their illness) both literally and metaphorically out of themselves (McCormack, 2004; Strang, 2004). These phenomenological acts of engagement between self and place were reflected in relational and emotional feelings of terror and panic but also of relief, respite and physical replenishment (Conradson, 2005a; Milligan and Bingley, 2007). The performances were also mobile with traditional movements around the Turkish bath, repeated in sequence from warm to hot to hotter and back to cold, mirrored in sweat-house practices of preparation, rest within the curative setting and follow-up acts of plunging and wrapping (Breathnach, 2004).

While the tepid springs at Mallow Spa were occasionally used for sweating-cures, the conflation of Turkish baths and sweat-houses is used to engage with wider narratives of embodied healing waters (Rutty, 1757). In contrasting two forms of sweat-cure that existed in the same country at roughly the same time, difference and diversity are evident in the cultural production of healing places (Williams, 2007). There is an opportunity also to contrast two types of therapeutic landscape, which were, in their cultural production, representative of the 'typical and unique' and the vernacular and colonial (Strang, 2004).

They exemplify similar processes of embodiment and cure in urban and rural settings, although the curative aspects of the later hydrotherapeutic treatments were of a significantly different form than those of the Turkish bath. But it was the role of the body as self-medicator, and its ability to cure itself, which was at the heart of these two places; where the simple pragmatism of the vernacular patient and the considered holism of the hydropath met on common discursive ground (Wear, 1992). Both forms functioned initially as places of informal health, though their meanings, practices and effectiveness were contested by developing systems of formal care in the later nineteenth century (Jones and Malcolm, 1999). This applied in particular to the more medicalized strands of hydrotherapy developed at one site in particular, Blarney. Finally the two types of 'sweating-places' shared commonalities of material structure (in the form of sealed heated rooms and plunge pools) and curative performance (via perspiration and wrapping) in their construction as therapeutic landscapes.

The first Turkish baths in Blarney were located in a tranquil rural setting, yet subsequent Turkish baths were constructed in the heart of towns and cities. They also contained elements of refuge, stillness and otherness, which contrasted with the busy streets outside. While less explicitly reported as therapeutic, the natural settings of sweat-houses also had qualities of stillness and remoteness (Burns, 1995; Conradson, 2007). Shifrin lists 49 Turkish bath establishments (of various types) in Ireland, while official Records have a count of 279 (234 in the South, 45 in the North) extant sweat-house sites in 2009 (Shifrin, 2009; DOEHLG, 2008; NIEA, 2009; see Figure 4.1). From these, three were chosen as representative examples. St. Ann's in Blarney, Co. Cork (see Box 4.1) was Barter's home and later developed into a hydropathic centre, while of nine different baths in Dublin, the Lincoln Place Baths are used to represent different phases of development (see Box 4.2). Tirkane sweat-house in Co. Derry (see Box 4.3) is used to illustrate the typical sweat-house. For the other sites, selected Turkish baths of Cork City are included as well as parallel Turkish bath and Hydro developments at Bray, Co. Wicklow and Monkstown, Co. Cork. The remaining sweat-house place-studies focus on clusters from individual counties including, Cavan, Leitrim, Roscommon and Louth (see Figure 4.1).

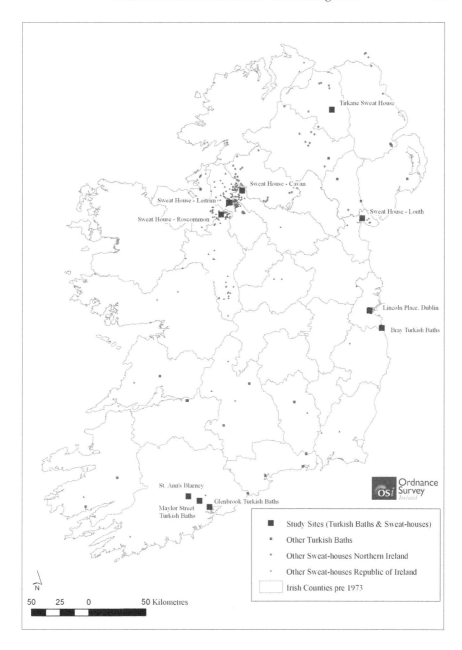

Legend:
- ■ Study Sites (Turkish Baths & Sweat-houses)
- ▪ Other Turkish Baths
- · Other Sweat-houses Northern Ireland
- · Other Sweat-houses Republic of Ireland
- ▭ Irish Counties pre 1973

Tirkane Sweat House
Sweat House - Cavan
Sweat House - Leitrim
Sweat House - Roscommon
Sweat House - Louth
Lincoln Place, Dublin
Bray Turkish Baths
St. Ann's Blarney
Glenbrook Turkish Baths
Maylor Street Turkish Baths

Ordnance Survey Ireland

50   25   0   50 Kilometres

**Figure 4.1 Map of the Distributions of Turkish baths and Sweat-houses with
10 place-study locations**

Source: Shifrin, 2009; Boundary Maps, Ordnance Survey Ireland. Sites and Monuments Records (Northern Ireland: NIEA, Republic of Ireland: DoEHLG). © Ordnance Survey Ireland/Government of Ireland Copyright Permit No. MP 008009.

Embodied Performances of Health: *Hammams* and Healing

Nineteenth century shifts from wider cultures of balneotherapy to a more focused interest in hydrotherapy, were reflected in bodily performances of health. While the basic symbolic understanding of water as therapeutic substance was unchanged, there were subtle variations in its meaning and production (Mackaman, 1998). In the original balneotherapy enactments of drinking and bathing, the minerals of the waters were natural elements given symbolic healing power (Gesler, 2003; Strang, 2004). With a shift to the more scientific and culturally-produced associations with the 'hydro', that symbolic healing dimension was given an altered identity. In both cases, health and healing were at the core of that meaning and function. In their common performances of curative sweating, both Turkish baths and sweat-houses echoed long-established folk-medicine techniques from a range of global cultures (Aaland, 1978; Wilson, 2003). The vernacular origins of the sweat-cure and its reproduction in social settings ranging from the *thermae* to the *hammam*, suggested a form of embodied curative memory inherent in societies both 'primitive' and 'developed' (Jackson, 1990; Porter, 1990). When seen in formal biomedical terms, the functions of sweating, in the regulation of body temperature, the sweating out of toxins, the treatment of skin and arthritic conditions and the building up of a range of immunities, offered a range of established health benefits (Berger and Rounds, 1998; Rojas Alba, 1996). However, in common with many 'therapeutic' processes, sweat-cures also had a health-endangering component, especially in relation to pregnancy, respiratory and cardiovascular systems, all of which were affected by excessive temperature and sweating (Berger and Rounds, 1998). A notional 'therapeutic potential' may better describe the healing properties and treatments of the 'sweating-places', given they were specific to individual physiologies and conditions (Conradson, 2005b; Williams, 2007).

The curative performances and outcomes at the two types of sweating-places were similar, given the physical processes involved. The most common treatments at the sweat-houses were for rheumatism, arthritis, fevers and backache. This was in part a function of geography, with many rural dwellers living in relatively cold and wet mountainous areas within poor-quality housing. In addition, agricultural work was physically demanding, hence the prevalence of arthritis and backache (Weir, 1989). At the Turkish baths, the treatment of specific illnesses included similar chronic conditions, especially rheumatism and arthritis. At St. Ann's, these included sciatica and scrofula. Broader muscular and bronchial complaints were also treated at the hydro, characteristic of many of the early sanatoria (Porter, 1999). Later treatments were linked to the new forms of mental health conditions associated with World War 1, especially in relation to rehabilitation and rest-cures for shell-shock (see Box 4.1). Finally, non-human health featured at both St. Ann's and at the Turkish baths at Lincoln Place in Dublin, where horses, cattle and smaller domestic animals were treated in a separate establishment with an entrance around the corner in Leinster Street (Shifrin, 2009).

Materially, the medical properties of both the Turkish bath and sweat-house were based on the physical production of places where intense heat, often over 200° F, led to changes in body temperature and profuse sweating. While the regulation of the temperature was relatively crude, even non-existent, in the sweat-house, the Turkish bath was based on a number of different models. In traditional and modern Arab cultures, *Hammams* generally consist of a series of open steam-heated rooms, which range from very hot to relatively cool. What Barter did was design a variant, partially based on studies of Roman *thermae*, named the 'Improved Turkish bath method', whereby the steam was extracted along vents in the floors and side walls via a large chimney. In addition, movement was between three rooms; in turn, cool, hot and hotter as well as a plunge pool (in his final version only) to intermittently regulate body temperature (Price, 1981; Shifrin, 2009). At a sweat-house such as Legeelan (Co. Cavan), typical practice was for a turf or wood fire to be lit several days before its use to heat up the stones (Richardson, 1939). Once heated, the stones were removed and rushes or bracken placed on the heated floor. The patient or patients crawled through a small door, which was sealed behind them and sweated for between 30 minutes to an hour and occasionally longer (Logan, 1981; Clancy et al., 2003). The structure of the treatment meant there was little variation in temperature, but after the treatment was over, most sweat-houses (as at Tirkane, see Figure 4.4) had a nearby stream or pool, which the patients used to cool themselves down (Weir, 1979).

As with other performances of health in place, the healing reputations of the 'sweating-places' were significant factors. As sites of very real, experienced and active encounters they reflected wider contestations around self and externally managed regimes of embodied health (Foucault, 1976). In both settings, this space between self-medication and medical supervision was sometimes filled by a range of agents with contested medical reputations. In a world of folk medicine and the associated steamy mystique of the East, there was plenty of room for new practitioners, some of who were seen as plain quacks or charlatans (Porter, 1989; Kelly, 2009). While the sweat-house can be considered to be a more independent and self-managed site of bodily cure, when compared to the more supervised setting of the Turkish bath and hydro, even here there was an informal supervision by friends and family, while itinerant 'bath-masters' were recorded in Louth and in parts of Cavan (Weir, 1989). Their role was to ensure that asthmatics and those with weak hearts or high blood-pressure were excluded and although their formal medical qualifications were non-existent, their experience and knowledge was trusted (Logan, 1981). At Tirkane, there was a record of patients being dragged out when they overheated, so such practices were not uncommon (see Box 4.3).

The body was central in the use of both baths and sweat-houses (Mackaman, 1998; Breathnach, 2004). Embodied water – derived from the sweat of the patient – became self-curative via the exclusion of toxins, loss of weight and the sweating out of fevers (Berger and Rounds, 1998; Frost, 2004). It provided an internally-owned form of healing that was, in a sense, 'inside out'; as opposed to other forms such as ingestion, which were 'outside in'. In the embodied process of the 'sweating-cure' it was the drawing out of one's own waters, of which the body was approximately 60% constituent (depending on gender and age) that provided the cure. In addition the ambiguous terms of sweat and steam reflected a contestable state of health and healing (Strang, 2004). It also had a strong pragmatic dimension. In various reports of the use of sweat-houses in Derry, Cork, Leitrim and Roscommon, as well as Rathlin Island off the coast of Antrim, all point to their therapeutic use for eye cures (Buckley, 1913; Weir, 1989). This health need was linked to the everyday experience of impoverished rural-dwellers living in smoky, steamy interiors of poor-quality housing (Mulcahy, 1891). There were also affective aspects of retreat – from harsh exteriors to interior spaces of calm and silence – in both the Turkish bath and the sweat-house. In the cocoon-like spaces of both, the body was warmed, wrapped, rested and safe in contrast to exterior experiences of toil, stress, cold and noise (Aaland, 1978)

As more complex hydrotherapies were introduced throughout the nineteenth century in places like Blarney, Monkstown and Lisdoonvarna, they reflected polarized positions of popular patronage and medical opposition (Mackaman, 1998; Shifrin, 2009). These contestations were two-fold; based partially on the perceived dangers of the sweating-cure as well as on how it reflected liminal cultural performances at the watering-place. The Turkish bath setting had an association which was health-derived, but had pleasurable aspects as well (Breathnach, 2004). While the baths were not seen, at least in explicitly biomedical terms, as curative, they had healing dimensions. In this they reflected reputational associations in both historic and contemporary spa settings of relaxation, rest and recuperation (Gesler, 1992; Kearns and Gesler, 1998). However these personalized restorative associations were not widely promoted in the Victorian world and it can be argued that pleasure and relaxation were only allowed when the contrapuntal elements of vigorous exercise and some sort of pain/suffering were experienced (Harley, 1990). Here the emphasis was more on the pill than the sugar, a reversal of the discursive productions of meaning at the spa but reminiscent of the privations of the health pilgrim and the presence of a moral geography of health at the watering-place (Brockliss, 1990; Bergoldt, 2006).

Box 4.1 St. Ann's Hydro, Blarney, County Cork

St. Ann's Hydro was located around three kilometres west of Blarney and was the creation of Dr. Richard Barter and subsequently run by his descendants. Barter had originally worked in Mallow and took out a long lease on the land and surrounding farm in 1836. His initial interest in the Turkish bath stemmed from meetings with Claridge and David Urquhart, British advocates of wider hydrotherapies developed by Priessnitz in Austrian Silesia. After visiting sites at Malvern and Ben Rhydding, Barter open his new hydro in 1843. His initial experiments were of a *bania*-like vapour bath but he eventually got the design for a new 'improved' form of dry-air bath right around 1858. After the initial development and construction of a small Turkish bath building, Barter developed the site by adding large new Turkish baths just three months before his death, in July 1870. After his death his son, also Richard Barter, managed the hydro and farm while employing a series of resident physicians between 1870 and 1916. The last of these, in 1911 was his nephew, Harry Barter. As a form of 'health farm', it was gradually extended over the years, so that at its peak it had about 80 bedrooms and in total, 110 apartments. Good fresh food was provided by the extensive gardens and farm up the hill behind the Hydro, which even had its own fish hatchery (O'Leary, 2000). The Hydro was at its peak from about 1880–1920 and then declined in the early decades of the new state. Harry Barter continued to be involved until 1940 and despite some investment by a new owner in 1944; the Hydro finally closed its doors in 1953.

Figure 4.2 St. Ann's Hydro, Blarney, Co. Cork. Panoramic View, c1914
Source: Courtesy of the National Library of Ireland.

The sheltered and wooded site, at an altitude of 250 feet, was considered therapeutically ideal. Initially it was a water-cure sanatorium but it was the 'Improved Turkish bath', which cemented the site's curative reputation. Developed as a franchise from 1859, Barter had considerable confidence in the design and was directly involved in its commodification and promotion outside of Blarney (Shifrin, 2009). In the period after the 1880s the site shifted its emphasis across to wider hydrotherapies. During the First World War, the Hydro was transformed into an auxiliary military hospital for wounded British and Irish soldiers. Many of the casualties arrived with prosthetic injuries and from this, the Hydro developed a genuine expertise in physiotherapy as well as new recuperative mental health treatments. Socially the site was patronized by a mixture of classes. While the Anglo-Irish and interested foreign visitors were the main clientele, Barter also made sure that a proportion of the rooms were made available to less wealthy clients, though some of these had to 'work their passage' to afford the relatively expensive cure. Facilities in the hydro included tennis courts, reading-rooms, billiard rooms and even an American bowling-alley. The site was further commodified with the construction of a railway station at the foot of the hill in 1888 while the social facilities included extensive woodland walks, fishing, a new golf course (built in 1907) and the tourist attractions of Blarney and Cork (O'Leary, 2000).

In terms of how Turkish baths and wider applications of hydrotherapy were received, there was considerable contestation from the medical establishment (Price, 1981). Originally developed in consultation with David Urquhart, the main British proponent of both hydrotherapy and the Turkish baths from the 1830s on, Barter's patented design, an improvement on Urquhart's more vaporous version, was considered downright dangerous by a number of medical opponents (O'Leary, 2000; Durie, 2006; Shifrin, 2009). This contestation was played out in spiteful interchanges and attacks in pamphlets and letters. A particular attack in 1861 by a Dr Gibson, directly aimed at Barter, but reflecting recurrent opinions around the dangers of high temperatures and vapour, and the need for the skilled and properly trained physician, noted:

> The Turkish bath has great attractions for lazy and luxurious people, with whom the 'killing of time' is an important consideration, but it would not be amiss for such people to inquire, whether a temperature of 150° does not press both heart and pulse to a gallop that will carry them to the end of life's journey sooner than they contemplated. Some learned physiologists assert that our span of existence is regulated by the number of our pulsations. Be this as it may, the heart that beats the fastest does not generally beat the longest. (O'Leary, 2000, 12)

However Barter himself wrote several considered responses to these charges and pointed out that while sudden changes in temperature and the presence of vapour were indeed medically dangerous, neither were practices carried out in his establishments and he should not be deemed guilty by association (Shifrin, 2009). It should be pointed out that these contestations worked both ways and that proponents of hydrotherapy were equally vocal in their critiques of scientific medicine and its 'drugging' practices (Price, 1981). These contestations of medical impact rarely extended to sweat-houses in part due to their 'native' positionality. Apart from a few 'picturesque' colonial commentaries, they were effectively dismissed as peasant curiosities, with any notion of their natural curative powers, or associated dangers, deemed unworthy of comment (Milligan, 1889; Mulcahy, 1891).

An interest in naturotherapy and 'health-as-naturalness' was a core component of the varied philosophies of hydrotherapy (Bergoldt, 2006). There was a certain tension however in how these interests in natural therapies were performed through invasive forms of treatment, at least in bodily terms. Both of the hydrotherapy establishments at Bray and Monkstown advertised a range of douches, massages and Priessnitz wraps; curative practices involving wrapping in several layers of wet blankets (Davies, 2007; Shifrin, 2009). But hydrotherapy was simultaneously part of a new range of natural therapies developed in the German-speaking world (Bergoldt, 2006). Here concerns with a more holistic model of health that combined cold-water treatments with careful monitoring of diet, exercise and even spiritual aspects of mental well-being were given practical and metaphoric meaning. This was partially reflected in the identity of the sweat-house as representative of a parallel folk-healing discourse. Both represented forms of complementary and alternative health approaches, which challenged the fast developing biomedical models.

In the settings of the Turkish bath and sweat-house, the wider enactments of an experiential health were clearly visible. As contrasting built/exotic and natural/ vernacular settings, both had affective and therapeutic components implicit in their curative spaces. These settings shaped experiential performances of health, where the whole body was both involved and immersed in place. Water's role was both internalized and externalized through that embodiment and via the wider use of water to cool the body down in a range of plunge baths, pools and streams. Yet water was also transformed by heating to assume new forms that could be both curative and dangerous, depending on the individual experiencing it. As expressions of wider forms of hydrotherapy, where a self-health dimension was enacted and performed by individuals, there were also tensions around behaviours and regulations. The two forms can also be seen as places wherein certain aspects of moral geography were enacted, around the body, health and hygiene, but also in issues of control, liminality and behaviour (Shields, 1991). As hydrotherapy became more mechanized, ironic given it original designation as an expressly natural form of therapy, some of its forms, in particular the Turkish bath, developed divergent medicalized and socialized cultural performances (Breathnach, 2004). These present a contrast to the original dimensions of therapeutic self-place encounters and it is to these contradictions that we now turn.

Cultural Performances of Health: Leisure and Pleasure

The two sweating places, though similar in embodied processes, had variations in their cultural performances of health (Lorimer, 2005). While both had curative dimensions, the sweat-house was, ironically, the most functional and health-specific setting of all those discussed. Although sweat-houses had a small social dimension, it was telling that it was at the more colonially produced health setting of the Turkish bath where a greater liminality and contestation of curative practice occurred. While the initial Turkish baths in Dublin, Cork, Bray and Monkstown were developed on the Barter model, their operational and cultural identities began to shift as wider colonial and class identities emerged (Breathnach, 2004). In particular Breathnach drew attention to the role of the Turkish baths at Blarney, Dublin and Bray, as settings in which colonial and native expressions of health, leisure and social class were played out against wider understandings of identity and authenticity:

> The Turkish baths established in Ireland during this period were found to be a site where several identities were expressed, including those of class, nation, and gender. The dual concerns of health and leisure were paramount to the formation of the baths. Firstly, the promotion of their redeeming qualities both physically and morally can be read as part of the move towards rational recreation, a movement that expressed a basic anxiety about the experience of pleasure. (Breathnach, 2004, 172)

Just as the development of the spa was an expression of the 'time freedom' of the gentry and upper classes, so the use of Turkish baths was an expression of a more commodified and egalitarian 'time constraint'. The social functions of the baths were as much a part of their meaning and material use as their curative reputations, and they became sites of leisure that represented the developing identity and confidence of the middle class, though with a strongly moral flavour (Breathnach, 2004). However there were also points of departure in how the baths developed the dual concerns of health and leisure. St. Ann's, and to a lesser extent the hydropathic establishments at Bray and Monkstown represented one strand of that development; private, curative, medicalized and supervised. The Turkish baths in Dublin and Cork, though still implicitly sold on cultural performances of health, therapeutics and hygiene, represented a more public, social and leisure-based strand (Shifrin, 2009).

In terms of class and colonialism, the Turkish bath had strong connections to Saidian notions of Orientalism and the 'Other' (Breathnach, 2004). These notions are especially relevant in the common displacement of local cultural forms by hegemonic colonial cultural identities (Said, 1985; Bayoumi and Rubin, 2000). A hybridization is visible in the cultural form of the Turkish bath and this reproduction of exotic/colonial place is an early example of the globalizing therapeutic landscapes (Hoyez, 2007b; Williams, 2007). In addition the cultural meaning of the exotic within the mundane, described in passing in Joyce's *Ulysses*, was also connected to a set of symbolic and interpreted meanings within the material environments (Joyce, 1997). The pointed minarets and dome of Lincoln Place Baths presented an incongruous but arresting sight in inner-city Dublin (see Figure 4.3) but also represented a complex cultural expression of health setting (Breathnach, 2004). Finally Barter himself cited an interest in creating a healing place, which invoked a combination of metaphors that were simultaneously, Roman, Turkish and Hibernian. That Hibernian identity was also suggested by Breathnach as more imperial than native (Breathnach, 2004).

Surprisingly similar themes informed debates around the origins and cultural production of the sweat-house as healing place (Wood-Martin, 1892; Evans, 1957; Weir, 1979). At first sight, they were materially reflective of their settings and locales; pragmatic, natural, and instinctive responses to health care needs in harsh and remote upland areas. It was suggested that the decline of the sweat-house paralleled the holy well, with the development of new primary care and dispensary systems in the latter half of the nineteenth century (Logan, 1981; Burns, 1995). But debates on the cultural meanings and origins of the sweat-house ranged from suggestions of Norse and Viking origins, to their construction by returning Irish visitors from America and Europe, bringing with them stories of sweat-lodges and saunas (Weir, 1979; O'Leary, 2000). While the notion of the 'imported exotic' thesis was, as can be seen with Barter, quite plausible, it seemed unlikely in the case of the Irish sweat-house and they do seem to have had indigenous origins. Strang suggests that universal human-environment interactions, built in an epigenetic way, may explain a combined nature-culture discovery of the healing power of place (Strang, 2004). In the echoes of the sweat-lodge, *bania, hamman* and *temazcalli*, the Irish sweat-house sits neatly as a culturally self-discovered curative form. Other uses of dry-stone corbelled buildings in Ireland in the form of monastic beehive huts and animal byres suggested that form may also have preceded function (Weir, 2009). Indeed some accounts suggested that in observing the benefits of animal cures, humans learned a lot and Barter was separately reputed to have tested his hot-air baths on his own animals while his first bath was a distinctive 'beehive-shaped thatched building' very reminiscent of a sweat-house (O'Leary, 2000: Shifrin, 2009).

Cultural performances of health at the Turkish bath were again important to Anglo-Irish identity, though less significant than at the spa or sea-bathing resort. All three types of therapeutic landscape had functions as centralized spatial meeting points, both rural and urban, for what can be described as 'virtual communities' or 'class colonies' of Anglo-Irish (Urry, 2002). Access to the baths became more available to the middle-class citizen once the numbers of baths increased, especially in Dublin and Cork as well as in Belfast, Limerick, Waterford and Ennis (Beirne, 2006; Shifrin, 2009). These discursive relationships between place, reputation and health/leisure were in many ways expressive of wider notions of communitas and habitus, enacted in both normative and recursive ways (Gesler and Kearns, 2002;

Atkinson et al., 2005). By contrast the social aspects of the sweat-house were much more localized and place-based. Here the practices of communal healing had communitas aspects, with individual sweat-houses such as Legeelan (Co. Cavan) or Kiladiskert (Co. Leitrim), used by groups of families from the surrounding townland or townlands. This communal identification with place was a further example of *dinnseanchas*, whereby associations between named place, rural identity and healing practice were expressed within local cultural performances and narratives. These were also expressions of more grounded spatial communities for whom the sweat-house, conceivable as another expression of a 'people's spa', enacted parallel processes of assembly, cure and social cohesion to those carried out at the Turkish bath.

The specific cultural performances of the original Turkish bath were broadly similar in different towns, though different rituals, practices and controls were associated with their later hydrotherapeutic identities. Within the Barter franchise the layout was generally the same. At Dublin's Lincoln Place, there were three core rooms. The first was the preparation room or *refridarium*, in which patients were dressed by attendants in robes and wooden clogs in preparation for moving into the second warm room, the *tepidarium*. Here the temperature was around 100° F, though the temperature was variable around the room to allow the patients to acclimatize against their own bodily capabilities. The final hot room was the *sudatorium*, where temperature were between 130° and 150° but could be raised as high as 200°F. A journalistic account from Lincoln Place from 1860 noted that there were separate chambers where those extreme temperatures could be experienced (Shifrin, 2009). There were also private rooms, plunge pools and cleansing basins. Overall the performances of healing were expressed in embodied acts of dressing, movement, necessary rest, immersion and cleansing. While these processes were supervised in Lincoln Place by exotically dressed attendants, they were also self-directed and inhabited by the patients themselves.

Performances were more informally managed, though not totally dissimilar at the sweat-house. At a number of Cavan sweat-houses, the practice and performance of the cure was spread across a number of days (Burns, 1995). It took most of the time to heat the sweat-house, often requiring up to five donkey-loads of turf, as at Corrakeeldrum; a not inconsiderable resource requirement (Richardson, 1939). Once the sweat-house was heated up, it was laid out with rushes or bracken on the floor, though if steam was required, water was also sprinkled on the floors (if stone) and walls. The number of people involved in the actual cure could range from two to six. Given the relative expense they were sometimes used by different groups one after the other (Burns, 1995). One person was always needed as a minder for the patient who in general either lay or sat on the floor or on piles of rushes (Evans, 1957). They were sometimes covered and sometimes not. In terms of clothing, naked bathing was not uncommon so in the moral climate of the times, they were sorted by gender. In typical performances, clothes were left in a pile at the entrances, though a commonly reported prank was for children to steal the clothes before the patients emerged (Burns, 1995). Once the patient had sweated as much as was deemed necessary, they either came out and used a stream, or as at Tirkane, a small plunge pool beside the sweat-house (Weir, 2009). They would then wrap themselves up as warmly as possible and head home to bed to complete the cure. Here there were shifting performances of health, which were similar in form if not in luxuriant cultural practice, to those enacted at the Turkish bath. It was likely to have been a purgative and even uncomfortable experience in the cold and inhospitable Irish climate and even 'out-victorianed' the Victorians as a performative act of pill without sugar.

The design of the Turkish bath and the operational use of the sweat-house both represented gendered spaces of health. All of Barter's original baths followed a broadly similar pattern where the different bathing rooms were split in two with parallel treatment rooms for men on the right and women on the left (Shifrin, 2009). Such separation was extended to the hydrotherapy treatments at St. Ann's, especially in the case of douches, given their occasionally invasive use of intimate body parts (Mackaman, 1998). The Turkish bath at Leinster Street was opened in 1882 as a male-only establishment though this was relatively short-lived as the bath's developers, the Sloanes, subsequently opened a second bath for both sexes at Number 3, St. Stephen's Green in 1888. At sweat-houses, genders also entered the baths separately and it was seen as the woman's role to tend the fires and keep them warm (Clancy et al., 2003). One peculiarly gendered use of sweat-houses was noted by a number of commentators on Rathlin Island. There were five or six sweat-houses on the island that were extensively used by younger woman before the Lammas Fair at Ballycastle (Mulcahy, 1891; Buckley, 1913). Here beauty was at the heart of the cultural performance as the sweat-houses were used along with healing varieties of seaweed, both of which were particularly effective in clearing up skin problems. In echoes of spa town performances, women wanted to look their best for any matrimonial opportunities that might present themselves (Buckley, 1913). Wider holistic and classical understandings of health as containing a beauty component are also worth noting as an oblique but for many, important aspect of well-being, a notion that is revisited in the setting of the contemporary spa (Heller, 2005; Bergholdt, 2008).

Box 4.2 Lincoln Place/Leinster Street Turkish Baths, Dublin

The Lincoln Place/Leinster Street Turkish bath was opened in February 1860, though the building had a previous life as a private medical school, originally run by Sir William Wilde (Fleetwood, 1993). The design of the building was by Richard Barter Junior, for the Turkish bath Company of Dublin Limited, set up in advance of construction in 1859. Richard Barter senior was still the major owner and it was the first example of his franchized approach. Early twentieth-century photos show an oriental appearance with two small minarets at the front street entrance, and a taller minaret and large 50' dome over the central part of the building (see Figure 4.3). There was a more prosaic but essential 85' tall brick chimney at the back. Internally, it followed the template of Richard Barter's baths at Blarney with three chambers, cool, warm and hot respectively, replicated for separate male and female baths and with separate entrances. The interior contained parallel sets of rooms for men and women and dressing rooms from which clients moved to a *tepidarium* and subsequently a *caldarium*, both with valued ceilings with variegated lights. There were separate small shower cubicles as well as a lounge area where 'bathers were able to rest after their bath and obtain coffee and a chibouk'. There was also an associated Turkish bath at the rear, with access via Leinster Street, specifically for 'Horses and other Animals' (Shifrin, 2009).

Figure 4.3 Lincoln Place Turkish Baths Shortly Before Demolition, 1960s
Source: Image from the photographic archive of Irish Heritage Giftware, info@ihpc.ie.

Additional changes were made in 1867 and 1875 that introduced more modern hydrotherapeutic features of plunge baths and showers. This shift in form was not unconcidental with the cessation of Barter's connection to the baths in October 1867. Competition amongst the main Dublin baths was intense and the baths went into temporary voluntary liquidation before being bought by the hoteliers who were their main competitors, Millar and Jury, in 1880. In the early 1880s new ventilation and heating systems were introduced though a high-profile court case for alleged negligence at the baths meant a tightening up of medicalized supervision at the baths. The baths were offered for final sale as a going concern in 1900, though no evidence of their operation as therapeutic site can be found after this date although the building itself was not demolished until 1970. Socially there was stratification in the earlier years of the baths, with separate first and second-class accommodations with differential pricing, both of which were considered relatively expensive for their time. As the competition intensified, prices and other social/leisure elements including a restaurant and accommodations were introduced. Performances at the bath were cited in both journalistic and imaginative accounts with the baths being identified as the probable setting for various liminal and suggestive episodes in Joyce's *Ulysses* (1997). Breathnach (2004) provides an intriguing account of the wider positionality of the Turkish bath in terms of social class and gendered identities with Victorian Dublin with evidence of an increasing middle-class patronage at sites like Lincoln Street.

While the leisure aspects have been touched on, the cultural performance at the Turkish bath, in common with other bathing settings, had deeper associations with liminality (Shields, 1991; Holloway, 2006). While detailed evidence of liminal behaviours are hard to come by, it was suggested in a number of accounts that they represented for both sexes, a place where conventional straightjackets, both physical and metaphorical, could be cast off (Breathnach, 2004). The casting off of corsets and waistcoats and the shared spaces of nakedness would have certainly appealed to many users, not all of whom would be exclusively interested in health and healing (Breathnach, 2004; Holloway, 2006). The illicit and occasionally scandalous public behaviours observed at baths and spas in a variety of earlier cultures, where sexes mixed in very free ways, were driven underground, or more specifically, indoors in the more controlled and moralistic Victorian era (Palmer, 1990; Kavita, 2002). Traditional eroticized associations with Orientalist and other public bathing spaces, were discursively understood through metaphors of *hammam*=seraglio/harem or the even older bath=bordello (Palmer, 1990). In hydrotherapeutic settings in France, the shift from public rooms to more private booths and smaller rooms led to concerns about homosexual activities and a moral as well as medical gaze became a characteristic of such places, while Breathnach likewise suggests an undercurrent of homo-eroticism in her analysis of gender roles at the Dublin baths (Mackaman, 1998; Breathnach, 2004). In this there are echoes of later contested performances of health and sexuality in gay bathhouses linking to their associations with the beginnings of the HIV/AIDS pandemic (Andrews and Holmes, 2007). In the liberated, othered and shadow (in a Jungian sense) performances of cure and leisure at the Turkish bath, one can see an imaginative connection to place, self and identity that was quite different from the more functional and pragmatic performances at the sweat-house.

In considering the cultural spaces of the Turkish bath, places like St. Ann's and other settings with literal 'sanitorial' functions, can be seen as forerunners to the modern spa, where traditional spa town practices were moved indoors and exoticized. Sweat-houses were also a type of secular spa, distinct from holy wells in that they were non-sacred but also representative of an ephemeral health response in particularly remote settings. Later colonial cultural performances shifted to newer hydropathic meanings and metaphors and in time to wider health-related meanings, especially around hygiene and rehabilitation. In addition, the Orientalist forms of the Turkish bath, an important initial public setting for hydrotherapeutic practices, aided this shift. In such exotic forms place was both materially and culturally re-produced in much the same way as contemporary spaces of yoga and wellness (Smith and Kelly, 2006; Hoyez, 2007b). Yet in a more locally exotic way, vernacular productions of healing hydrotherapies were also visible in the development of the sweat-house, which emerged from more local impulses and cultures (Logan, 1981).

Spatial Performances of Health: Steam and Smoke

In considering spatial performances of health, sweating-cure locations were to be found in both countryside and city (Gesler, 1992; Philo and Parr, 2003). While the original Turkish bath site, St. Ann's, was rural, the development of Barter's franchise after 1859 was primarily urban-based. New Turkish baths, often with Barter's direct involvement, were built within five years in Bray, Dublin (Lincoln Place) and Cork (Grenville Place and Maylor Street) and, in partnership with Urquhart, in London and Manchester (Shifrin, 2009). Subsequent developments were primarily in cities and large towns (Belfast, Waterford, Limerick, Ennis, Killarney, Kilkenny and Sligo) as well as smaller towns (Downpatrick, Dungannon, Newry, and Tipperary) and intriguingly, sea-resorts (Tramore and Youghal). The more rural location of sweat-houses and their extant distribution are marked in Figure 4.1. While they were most prominent in the border zone of Leitrim, Fermanagh and Cavan they were also to be found in marginal rural parts of counties Roscommon, Antrim, Louth, Derry, Tyrone and Galway and occasionally in other counties (Weir, 1979). It has been suggested that the decline of the sweat-house after the famine had a spatial association with the cultivation and inhabitation of marginal agricultural land (Weir, 1979). As the Famine enacted a form of 'Malthusian spatial cleansing', so populations literally retreated back down the mountain to be closer to more cultivated territory and by extension, closer to newer sites of formal health care.

Box 4.3 Tirkane Sweat-house, Co. Derry/Londonderry

One of five extant sweat-houses in the county, the sweat-house at Tirkane, known in Irish as *Teach Allais Thír Chiana,* lies three kilometres west of Maghera. It is the only sweat-house under statutory management, in this case by the Northern Ireland Environment Agency. It is well signposted and there is a paved path running around 200 metres to steps that lead down to the secluded site itself. Here the sweat-house is set back against a bank where the stone construction, similar to an igloo, is slightly overgrown and covered over by grass and sods. The low entrance doorway is visible along with a hole in the roof. About 10 metres south of the sweat-house lies a small plunge-pool and rivulet leading to a small stream, while the whole site is set within a wooded dell with aesthetic qualities of seclusion, stillness and therapeutic affect.

Figure 4.4 Tirkane Sweat House, Co. Derry, 2008

The specifics of the sweat-house's material structure and operation are described on a formal notice-board beside the structure and neatly summarize built forms and healing performances, equally applicable to other sweat-house sites:

> The sweat-house has stone walls and a paved floor. The roof is lintelled using flat stones with a small air-hole in one corner. The structure is covered in turf and looks like a little grassy mound except for the entrance, which could be mistaken for the entrance to a badger's set. It is a mere 40cm square. The space inside is 1m wide, less than 2.5m long and about 1.7m high. It may seem tiny, but as sweat-houses go, this is about average, big enough for four or more people.

A large amount of fuel was required to get the sweat-house thoroughly heated. A big turf or wood fire was lit. When the interior was very hot, the fire was drawn out, the floor swept, and green rushes or bracken strewn thickly over it. Sometimes water would be thrown over the floor to create steam. The patients undressed and crawled in. The entrance was closed up; the air-hole in the roof was covered and the inmates were left to sweat it out in the dark. The air-hole could be uncovered to adjust he temperature if necessary. Patients who were infirm or already weak often fainted from the heat and had to be pulled out. After an appropriate time, about half and hour, the entrance was unblocked'. (NIEA, 2008)

The fact that the site is preserved and managed as a tourism and heritage site is emblematic of a wider potential commodification of healing place, though here it is primarily a reminder of former practice. Yet the notice-boards at the site also provide commentary on sweat-house origins and also note the specific curative power of this example of therapeutic landscape: 'A cross between a Finnish sauna and a Turkish bath, the Irish sweat-house was considered highly effective in treating rheumatism and all sorts of aches and pains' (NIEA, 2008).

Given that most of the Turkish baths were in urban locations and close to trains or trams, they were relatively accessible to their users. Indeed the three Dublin baths were right in the heart of the city along its main street (Sackville (now O'Connell) Street) and beside Trinity College (Lincoln Place/Leinster Street) respectively. St. Ann's was initially five kilometres from Blarney station, while a subsequent station on the Cork and Muskerry Light Railway was only five minutes walk; built on land specifically donated by the Barters for the purposes of improved accessibility (O'Leary, 2000). Boats as modes of transport reappeared at Monkstown with new piers being built for steamers from Cork city in 1859, while the site was connected by rail shortly after (O'Mahoney, 1986). Sweat-houses on the other hand were quite different in terms of both access and utilization. Here the remote therapeutic settings meant that access was almost invariably on foot with horses and carts used only for the transport of fuel (Burns, 1995; Gesler, 2003).

Clienteles at St. Ann's were undoubtedly from the 'better' classes of the Anglo-Irish and included a consistent patronage from Britain and beyond, with recorded visitors from France, Germany and America (Sheehan, 1995). Elsewhere, the clienteles was broadly similar to those of the spa town, with the gradual embourgeoisement of the Catholic middle class from the 1880s on, noted by commentators on the Turkish bath (O'Leary, 2000; Breathnach, 2004). In part this was a purely commercial process, as the baths struggled to sustain business and opened up to a range of previously unconsidered users such as commercial travellers and other tourists. This was evident in the 4.30 a.m. opening hours at Lincoln Place, with bath and breakfast provided to clients en route to the morning ferry to Holyhead (Shifrin, 2009). These new locations of therapeutic leisure provided opportunities in new and even anonymous settings for previously limited social mixing. Breathnach notes the relationship between dress, undress and social class in the fact that for middle-class Victorians, the way one dressed marked one's social position and this hierarchy was sometimes lost in the near nakedness of the baths (Breathnach, 2004). At the Lincoln Place or Sackville Street baths, as in the French spa town, the opportunity for spatial, social and sexual mobilities were enacted through previously excluded contact at the watering-place (Shields, 1991; Mackaman, 1998). At the sweat-house any notions of exclusivity was easily dispelled in the pragmatic and functional rural utilization of the sites and its communal operation by families and neighbours.

Though the Turkish bath could be considered an elitist therapeutic landscape, it is worth recording a separate aspect of the roles of Barter and Urquhart in its popularization. Both men were broadly socialist in their leanings and the initial urban bath developed by Barter at Maylor Street in Cork was explicitly named, 'The Improved Turkish or Roman Baths for the POOR' (O'Leary, 2000; Shifrin, 2009). This operated from 1863 to 1894 and was paid for from voluntary contributions and grants from the 'Barter Bath Charity Fund'. Indeed both Barter and Urquhart were on record as believing that baths should be affordable to all and seem to have been as good as their word. However, there was a certain moral conditionality attached and Barter himself stated:

> I have erected a complete bath for the poor at my own expense, and they resort to it in hundreds; and see what a boon this is to them! It removes from them the necessity for stimulant drinks; and I venture to affirm that it is the means best calculated to take from Ireland its two great evils – intemperance and filth. (Barter, 1861, cited in Breathnach, 2004, 164)

Though Barter planned baths for the poor in all the major cities of Ireland, the only other one completed was in Belfast (Shifrin, 2009). The associations with hygiene and temperance suggested that while curative forms of health were to be found at the hydro or at Turkish baths like Ennis, the poor needed a separate cleansing and moral surveillance so for them the bath water was deemed to have a differential identity and function (Breathnach, 2004; Beirne, 2006).

In the case of the sweat-houses, seasonality was reflected in their most intensive use coinciding with late summer and autumn. This was the case for Cavan, though in Leitrim, seasons lasting from April to September were noted; the winter and early spring were generally considered too harsh for operating what was essentially an outdoor cure (Weir, 1979; Burns, 1995). Different aspects of temporality informed accessibility and utilization at the Turkish baths. As indoor sites, the spas of Dublin and Cork had year-round seasons but they were, in contrast to the more residential hydrotherapeutic settings, utilized in a much more short-stay form. This distinction in terms of length-of-visit was also a factor in the social stratifications and clienteles that attended the baths. This must however be tempered by wider 'time-constraints' associated with newly-industrialized societies. While urban baths were not cheap, they were more accessible to a busy and increasingly middle-class clientele who did not have the time to linger for a longer visit (Mackaman, 1998; Breathnach, 2004).

While other therapeutic landscapes had a longer time-span, there was a more marked period of development and use for both Turkish baths and sweat-houses. Unlike the *temazcalli* of Mexico, which are still used today, the sweat-houses went out of meaningful utilization as therapeutic places at the end of the nineteenth century, though occasional later uses were recorded (Weir, 1989; Rojas Alba, 1996). Their construction was relatively simple and so ephemerality and temporality were undoubtedly common as local populations waxed and waned. The Turkish baths required a more considered and resource-intensive development, which were often astonishingly short-lived. Shifrin's excellent and authoritative web-site records the specific dates of opening, operation, ownership and probable closure, which identify in detail the ephemerality of these types of therapeutic landscape (Shifrin, 2009). Bray for example, was opened in a blaze of publicity in November 1859, yet within five years was looking for new owners and was sold off by 1870 (Davies, 2007). Ennis Turkish baths lasted from 1870 to 1878 while others barely lasted a year. In contrast some of the Dublin baths had a longer history. Sackville Street survived in various forms from 1869 until 1922, when it had the misfortune to be used as a hide-out by senior leaders of the Republican side in the opening days of the Irish Civil War and was bombed to smithereens (Boydell, 1984).

The therapeutic properties of place were also important in the settings of Turkish baths and sweat-houses (Kearns and Gesler, 1998; Conradson, 2007). Here a more three-dimensional and inhabitable version of landscape was invoked in the making of healthy place (Gesler, 2003; Williams, 2007). St. Ann's was described, and indeed to an extent sold, by explicitly linking therapeutic associations of place and human health, as it was cleverly built on the lee side of a wooded valley to create shelter and a micro-climate to suit the health identity it was to assume (O'Leary, 2000). Surrounding woods and parklands were extensively used in the performances of the health cure while the large sun-lounge was a forerunner of later therapeutic design elements as well as a necessary diversion in the all-too regular inclement Irish weather (O'Leary, 2000; Curtis, 2004). The Baths and Hydropathic Establishment at Monkstown stood overlooking a narrow section of the River Lee as it flowed out to Cork Harbour and its riverine setting was described as 'commanding a prospect unsurpassed for beauty' (O'Mahoney, 1986). Tirkane Sweat-house had a secluded location on the edge of the Sperrin Mountains, described in modern tourist brochures as 'set in to the side of a small leafy valley' which opened out into a natural space complete with small stream and plunge pool (Sperrin Tourism, 2008). Almost all sweat-houses were to be found in remote and often secret locations, in part due to their functions as private spaces of care but also, according to Weir, on the need to hide them from the local landlord, who might demand ownership or a tax on their utilization (Weir, 1979).

The built environments of Turkish baths were also explicit reproductions of 'other' and visibly Orientalist spaces (Said, 1985; Hoyez, 2007a). The Turkish baths at Lincoln Place in Dublin had exteriors containing turrets, small minarets and onion domes. Inside the baths, that impression was augmented with Moorish-style interior decorations. This decoration was found on doorways, arches and panels as well as on the tile work, which lined the wall. Sometimes, the exteriors were more exotic than the interiors, which were plainer wood and brick constructions as at Blarney, Maylor Street and the later Dublin baths. The 'effect-affect' of the built environment was then topped off with more inhabited and ritualistic aspects of costume, hookah and coffee. In contrast, though no less affective, sweat-houses looked like small stone igloos with low entrances and small holes in the roofs. Because they used local materials, primarily stones, they reflected their local setting and were often overgrown and layered with sods and grass to both disguise and insulate them. But they were, in contrast to the more artificially made Turkish bath, invocative of a more natural and vernacular construction. The connections between the two were noted however in Barter's first attempts and in an Irish Builder article about the Turkish baths of Dublin where they were described as: 'what is in reality a revival and more perfect development of the old Irish *Tigh Fallais* (Sweat-house)' (Breathnach, 2004, 169). Ironically, reverse invocations were noted for sweat-houses where; 'The hot-air bath, or what would now be designated the Turkish-bath, itself but a degenerate imitation of the luxurious *laconicum* of Ancient Greece and Imperial Rome – was in common use amongst the Irish' (Wood-Martin, 1892, 411).

In considering the broad similarities of historical health narratives, practices and meanings of the sweat-cure, mobile rehabilitative and restorative meanings were physically enacted within therapeutic place (Gesler, 1993; Williams, 2007). Within the natural and built settings and locations of the Turkish bath and sweat-houses, the therapeutic engagement between self and landscape, patient and place was culturally formed and individually experienced (Conradson, 2005b; Milligan and Bingley, 2007). In the steamy surrounds of stone igloo or tiled room, the setting was part of that affective process. A visitor, describing himself as 'A Moist Man' provided an 1860 account of the Lincoln Place baths, which noted a range of affective objects within the different rooms:

> ... I was conducted into a large room around which were arranged little curtained pavilions about the size and shape of a four-post bedstead. The room was decidedly Turkish in its aspect and appointments, the crescent form being as far as possible given to every thing, while ottomans and other matters of oriental furniture were to be seen. The servants wore long scarlet flowing dressing gowns and Turkish slippers, and on stands were arranged trays with china coffee-cups, etc ... the calidarium ... was still more ottomanic than the [tepidarium], being in dim prismatic twilight, and without windows, unless the little star-shaped scraps of crimson, blue, and amethyst-stained glass, artistically inserted in the vaulted roof, could be so-called. (Shifrin, 2009)

In this setting, exotic elements were deliberately used to enhance an affective mood and emotion, already heightened by the passage from a busy and bright Dublin street to a quite literal 'other' world. This process was also likely to create the conditions for a more embodied and self-expressive understanding of the other, and by extension, self in place (Sibley, 2003; Breathnach, 2004). Breathnach also cites Inglis in this liminal link between the body and feeling in a world where neither were encouraged and where the 'othered' experience was; 'a luscious contradiction, lived in the body ... (where) you might become someone else, someone freed from propriety' (Ingold, 2000; Breathnach, 2004). This acculturation of the senses to water is therefore present in both healing and pleasure and in all of the embodied acts of ingestion, immersion, expulsion and introduction (Strang, 2004).

Economic Performances of Health: Franchise and Fashion

As settings for economic performances of health, there were significant differences between the two types of sweating-place. Turkish baths had a strong commercial focus, in stark contrast to the essentially free and communally managed sweat-houses. In the case of the Turkish bath, the central entrepreneurial role of Barter, and his relative wealth, meant that he could experiment with his 'improved' format, including sending his nephew to Rome to study traditional *thermae* (O'Leary, 2000). Here a combination of capital investment, time and scientific experimentation were necessary antecedents to entrepreneurship. In addition, Barter's strong commercial instincts, evident in the patenting, franchising and defence of his 'Improved Turkish bath', meant he was the type of capitalist entrepreneur particularly well suited to the Victorian colonial economy (Breathnach, 2004). His philanthropic side was not a contradiction, but rather a common combination for his time, though also enabled by the advantages of class, wealth and social position. While his Turkish baths were undoubtedly popular, they were founded along colonial narratives, both in their identities of exotic 'otherness' and in their utilization by the upper classes. While his dedication to the poor was laudable, it was also conditional and partial in moral and financial terms. At an individual level, Barter's descendants' philanthropic instincts did allow them to provide a low-cost option at St. Ann's. While the level of treatment and service for the second class patient was comparable to that of the better class of patient, there was a further class-based twist in the tale. In return for full medical care at a reduced rate, the second class patient was expected to work at the spa, typically as attendants or waiters (O'Leary, 2000). Here the positionalities of class and social stratification were recursively produced right through to the 1920's.

Barter's franchise model, where the management of the baths in Lincoln Place, Leinster Street and Bray was left to partners and major investors, was an effective way to keep a design and share-holding interest, while leavening the major risk on to the franchisee (Shifrin, 2009). That so many were relative unsuccessful commercially, suggest that Barter's grasp of entrepreneurial risk was a sound one. Nonetheless, as was noted previously, some of the city baths in Dublin and Cork were relatively successful in commercial terms. They were attractive to the burgeoning middle classes of the mid nineteenth century and were well patronized with 115,000 visits recorded at Lincoln Place Baths between 1859 and 1863 (Breathnach, 2004). But the marketing was not always domestic and an advertisement in Guy's Postal Directory of 1886 noted that the Blarney baths were better known in Germany than in Britain, a sign of local decline as much as international fame (O'Leary, 2000).

From a health perspective, entrepreneurship was reflected in the original Turkish and later hydropathic identities of the chosen sites. A strong business ethic sustained admittedly more 'mainstream' alterations to Turkish baths. At Lincoln Place, crystal fountains were built in 1867 to add moisture (a direct departure from the Barterian template), with modern showers and plunge baths added in 1875. A response to consumer demand was noted at Leinster Street, where 'continental' sprays had been installed in the shampooing rooms, 'at the request of gentlemen who visit the Vichy baths' (Shifrin, 2009). Hydropathic forms were most fully developed at Blarney, Monkstown and Bray and, in the latter two places, were spatially separated from the original bath sites. Hydros like St. Ann's were kept as up to date as possible with new healing therapies, especially massages and douches. Its health history reflected then-current health technologies as simple baths and showers gave way to douches from the 1880s and newer electrical and 'ergotherapy' forms in the early decades of the twentieth century (O'Leary, 2000). For Barter's descendents, St. Ann's tapped into the increased interest in hydrotherapies, an early example of the money to be made from different forms of CAM in Ireland. It also coincided with the development of a more hydropathic flavour at spa towns and sea-bathing resorts (Beirne, 2006).

More widely, the roles of reputation and fashionability reflected the global 'craze' for Turkish baths from the 1860s to the 1880s (Towner, 1996; Borsay, 2000). Exotic place invocations emphasized the reputational and symbolic value of the therapeutic performance, simultaneously aimed at the acculturated tastes of the Anglo-Irish (Borsay, 2000; Gesler, 2003). Barter's 'Improved Turkish bath' associated itself directly with Priessnitz, producing a promotional pamphlet called, '*Descriptive Notice of the Rise and Progress of the Irish Graffenberg, St. Ann's Hill*' (Barter, 1861). The more hygienic and pleasurable aspects of the urban Turkish bath also signified a general move towards commercialized leisure and the social side of bathing that reflected wider health/leisure shifts towards the sea (Corbin, 1994; Urry, 2002; Breathnach, 2004). By contrast, there was virtually no entrepreneurial intent in sweat-houses, given they were mostly built for local consumption. There are suggestions that they were shared or lent out to friends, neighbours and family. Whether this was on any sort of paid basis is unlikely, though there are records of one family building their own sweat-house as they did not like crossing neighbours fields to reach the nearest one (Burns, 1995).

Reflecting the competitive selling used at spa towns, health as a business was evident at the bath and hydro (Hamlin, 1990). Advertising and marketing were as important in the Victorian world as they are now. In Ireland this was reflected in the competitive press adverts used to promote newer competitors in Dublin against the original Barter baths. The Sackville Street baths were set up in 1869 and used the evocative company name, 'The Hammam Hotel and Turkish baths', complete with a star and crescent logo. By 1875 competition between the Hammam, Lincoln Place and newly opened Turkish baths at St. Stephens Green was played out through newspaper adverts and price-wars. Lincoln Place suffered as it was the most expensive, while St. Stephen's Green baths had access to more modern technology and facilities These combinations of reputation and cost were important and Lincoln Place's loss of its connection to Barter suggested that the conflation of place, person and symbolic reputation remained crucial in the successful maintenance of a therapeutic place (Shifrin, 2009). At Monkstown, the local press were also used in the competitive marketing of place. There was intense competition between the original Glenbrook Victoria Hotel and Baths owned by Robert Watkin Jones and the competing and neighbouring Dr. Curtin's Hydropathic Establishment. In early 1858, Jones reopened his Hotel and advertised the imminent opening of his new Turkish baths, followed by a grand dinner at which Richard Barter was to attend. On seeing this, Curtin let it be known in a competing advert that he too was about to open a new Turkish bath. Shifrin notes:

> Whether this was ever a real possibility (which was in the event delayed), or an acceptable commercial exaggeration, we may never know; the description of Curtin's forthcoming bath was clearly (what we should now call) a spoiler designed to diminish the impact of Jones's publicity. (Shifrin, 2009)

In addition, there is a suggestion that in a competitive world of advertising there was an increased use of visual imagery of relaxation and semi-clad men and women. This was both daring and seductive in the buttoned-down Victorian world and further emphasized the liminality as well as the healing/pleasure dichotomy of the typical therapeutic space, a fact noted in sites of spiritual healing such as séances, popular in the same period (Breathnach, 2004; Holloway, 2006). The increasingly difficult markets for the Turkish bath toward the end of the century also led to a subtle shift from health, to more beauty oriented aspects of shampooing, hairdressing and the sale of products; another fore-runner of the contemporary spa and wellness space (Shifrin, 2009).

In their economic performances, the therapeutic landscapes of the Turkish bath and Sweat-house reflected wider health-place production. This was particularly true at places like Bray, where the Turkish bath was but one health element in the broader development of a sea-bathing resort. The bath itself survived a mere six years from its effusive opening in 1859, while the Bray Hydropathic Establishment built across the road was equally short-lived, lasting only from 1858 to 1861 (Clare, 1998; Davies, 2007). A local entrepreneur, William Dargan, was centrally involved in both ventures, though in a sense they were 'loss-leaders' to a wider commodification of the town as a health resort (Davies, 1993). Discursive geographical connections were also reflected in more specifically Orientalist comparisons at Monkstown. Here exact geographical parallels were drawn based on the riverine setting and built environment; 'The Hotel and Baths of Glenbrook stand midway between Passage and Monkstown. Viewed from the river, they remind the traveller of a Turkish temple on the Bosphorus' (Gibson, 1861, 406).

Finally the use of imaginative fictionalized accounts, in common with the spa novels of the French watering-place, were used as thinly disguised marketing tools in the Cork city baths (Mackaman, 1998; Shifrin, 2009). Here fictionalized short stories in tracts were produced anonymously by a Mrs. Donovan, a member of the Maylor Baths management committee. In a tract titled, *Chat upon health: Pat Dennehy visits Mrs. Magrath,* a two-way conversation was written up with specific advice on health regimes, benefits and behaviours.[1] As a comment on the use of narrative and reputational accounts of the benefits of a therapeutic landscape, as well as a cleverly reflexive allaying of health fears, such artificial but suggestive exchanges could not be bettered, even if they could not always be believed (Gesler and Kearns, 2002).

Despite their seeming ephemerality, Turkish baths had an underestimated longetivity as a therapeutic landscape form in Ireland and indeed elsewhere. Ephemerality is by its nature a relative term and there was a continuous presence of Turkish baths in the city of Dublin from 1858 to the 1930s while the Hydro at St. Ann's in Blarney originally opened its doors in 1842 and did not close them finally until 1953 (O'Leary, 2000). Material elements linger in the lost landscapes settings of Dublin and the South Mall of Cork. At the latter, which is currently a restaurant, the faded name can be seen in the paintwork while in O'Connell Street in Dublin, a rebuilt Moorish doorway still marks the entrance to the Hammam Buildings. For the sweat-houses, notions of spatial and commercial competition were less

---

1   Pat – So you take the Turkish bath! People tell me I would be roasted there. What is the use of all this heat?

Mrs. McGrath – What is the use of the sun? One of the plainest marks of good health is a feeling of comfortable warmth, as cold is always characteristic of death. Ask the poor rheumatic cripple what good heat does him? Or the man who was comforted in the Bath in the first chill of illness. If the poor creatures we see on a winter's day shuddering with cold, at the doors of our Dispensaries, were put into the Bath, they would think themselves transported into Heaven.

Pat – If the Bath be so good, why isn't it in all the hospitals?

Mrs. McGrath – I leave that to others to answer. All I can tell you, Pat, is that there is no Institution in the City so valuable to you and me as the Bath. It not only washes the skin, but it also washes the blood. It fortifies against cold. It brings the blood to the surface of the whole body, as the hot stupe [poultice] does to a particular part. Its general use would change the whole condition of society, lessen the Poor Rate, and prevent cholera, small-pox, and similar evils, which all spring from dirt ...' (Shifrin, 2009).

important than simple local geographies of land ownership and the lack of choice
for health treatments. The lingering presences of the sweat-house are hard to find,
in part due to their specific construction in inaccessible locations as well as the
organic re-merging of cultural health form back to natural landscape, though some
specific examples are preserved and even re-commodified. Heritage and tourism
uses noted at the holy well are being re-enacted as 'CAM attractions'. While Tirkane
is fully managed by the NIEA, sweat-houses have been reconstructed as 'heritage
attractions' at Keadue, County Roscommon and at Parkes Castle on the shores of
Lough Gill in County Sligo. At the latter, the presence of an accessible example of
vernacular healing place at a major contemporary tourist site is reminiscent of the
commodification of St. Brigid's Well at Liscannor. However the revival of a dead
therapeutic space may be more attractive than potential damage to a living one.

Elements of political economy, power and ownership were as visible at
the Turkish bath as elsewhere. Power-relations were directly reflected in their
positioning as twin representations of alternative health and liminal leisure,
which acted as challenges to health and society respectively. While a number
of commentators noted concerns with the sensuous and affective enactments at
the Turkish bath these aspects were less contested than the attacks of the new
medical establishment on the dubious efficacy of the baths. Commentators such
as Dunlop were regularly engaged, especially at the start of the 1860s, in disputes
with Urquhart and Barter, while later attacks were aimed at proprietors like Wilson
at Malvern (Price, 1981; Durie, 2003b). Any evidence or account of death or
negative effects emerging from the baths was seized upon by both the medical
and mainstream presses. While Barter, as noted above, was effective in fighting
his own corner, the less scrupulously run and managed baths were more open to
contestation. There was a lengthy and reputationally damaging court case around a
case of carbon monoxide poisoning at Lincoln Place baths in 1881 (Shifrin, 2009).
The outcome of this and other early cases with iatrogenetic elements marked the
beginnings of a health and safety culture, relatively unknown in Ireland up to this
point (Porter, 1999). It also marked the beginnings of a managerial control expressed
in therapeutic settings through increasingly bureaucratic cultural dimensions of
regulation, qualification and professional, rather than self-ownership (Mackaman,
1999; Durie, 2003a). For the sweat-houses, such issues of power and ownership
were less important though they still existed in indirect ways. They represented
marginal therapeutic geographies where colonial processes of plantation and estate
enclosure had pushed native inhabitants to effectively invent their own therapeutic
forms from necessity and exclusion (Duffy, 2007).

Summary: Place and Perspiration

For more than 80 years, these (Hydropathic) methods of treatment have been
carried out at St. Ann's Hall, and, although the means have varied, for now

Electricity is used as a source of heat, the principles remain the same. (O'Leary, 2000, 26)

In this comment from a twentieth-century visitor to Blarney, historical shifts in health meaning in place were represented in St. Ann's mobile identity from holy well and Turkish bath to hydro and asylum. The shifts in function, form and meaning, especially of the Turkish bath and its augmented hydropathic expressions, were also indicative of spatial and medical mobilities. In adapting to wider fashions and meanings of health, where class, colonialism, habitus and identity were enacted, change were less to do with specific meanings of healing that with its expression. Similar classical connections had their apogee for Barter when a group of visitors from Baden-Baden, came to visit Blarney to get permission to bring back the patented method to their own hydro (O'Leary, 2000; Shifrin, 2009). There they set up what they termed a 'Römische-Irische Bad' and these hybrid Roman-Irish baths can be found in Wiesbaden and other continental spas to this day. In this nomenclature there is a literally correct but spatially hybridized conflation of healing narratives, authentic forms and inhabited settings characteristic of therapeutic landscapes more generally (Hoyez, 2007b).

Though less visible relics in the landscape than the spa town, the Turkish baths and sweat-houses were important elements in the space-time development of the Irish therapeutic landscape. The sweat-house was never a therapeutic form well-known beyond those regions of rural Ireland in which it was found (Weir, 1989). Yet in local histories from parts of rural Leitrim and Fermanagh, vernacular knowledge and traditional narratives of health and healing are recorded and stored, embedding the locations, conditions and practices of sweat-houses within national inventories and mirroring the cures of the holy well (NFC, 1934a; Clancy et al., 2003). In the simple stone and earthen constructions, produced by bodily labour and in traditional forms, natural and cultural elements of health in place are sustained in the rural landscape. While symbolic elements of escape, retreat and even a liminal sensuality were all bound up in the external promise of the Turkish bath, its symbolic derivation and internal decoration were a hybrid mix of Classical and Arab design with a functional Victorian underlay of brick and tile (Breathnach, 2004). As a relatively successful example of what was often an ephemeral business, the sustenance of St. Ann's Hydro was a separate testament to shrewd understandings of the paradigms and commodifications of health. In being a rare example of an Irish site of 'health production', the development of the 'Improved Turkish bath' franchise was prescient and an uncommon reversal of more typical colonial directions of place appropriation and reproduction. While an indirect role for political economy can be seen in the 'necessitive' spatial production of the sweat-house within a marginalized and poor rural community, it can also be seen as an un-commodified production of health, which like the holy well, stands as a vernacular health counterpoint to the more colonial productions of the formal watering-place.

In more specifically hydropathic settings, the different forms and designs of associated treatments were representative of medicine, science and progress. This avowedly modernist symbolism was built into the material practices at St. Ann's, where a new 'electrotherapeutic' department was developed in the early decades of the twentieth century for the treatment of neuro-muscular and arthritic diseases; later augmented by 'inductothermy' and ultra-violet and infra-red radiation (O'Leary, 2000). In generically naming all of these as 'ergotherapy' practice, Richard Barter (Jnr.) sustained the exotic metaphor but also marked a shift from CAM to a more scientific and technological medicine, ironic given the original holistic intentions of hydropathy (Bergoldt, 2008). In addition, the material aspects of the cure setting, whether in the exotically coloured and well-lit *tepidarium* or the dark, almost womb-like envelopment of the sweat-house, were designed for their affective healing properties (Holloway, 2006; Tolia-Kelly, 2006). In terms of material connections to other therapeutic landscapes, the Turkish bath was a logical development from the spa town through more sophisticated understandings of hydrotherapy while the sweat-houses can be seen as a domestic reproduction of hydrotherapeutic practice with loose links back to the holy well.

Beyond those material aspects of sweating-places, the established importance of metaphor in the production of such places was equally evident (Gesler, 1993; Geores, 1998; Williams, 1999a). Discursive links between class tastes and fashionability was reflected in their health behaviours with the aspirational habitus of the Anglo-Irish reflected in the exoticism of the Turkish bath. By contrast, the earthier habitus of the rural Irish was evident in the pragmatic and natural elements of the sweat-house. Metaphorical connections with 'health fashionability' were also evident in the developments of both Turkish bath and hydrotherapy, reflecting wider tendencies towards health 'crazes', though sweat-houses can also be seen as fashionable in a different way (Burns, 1995). Ephemeral relationships between health practices, class snobbery, human curative need and exploratory healing curiosities were also evident in these early manifestations of complementary and alternative therapies (Heller, 2005; Bergoldt, 2006). As with similar health 'crazes', both historic and contemporary, the curative efficacy and healing performances of these new forms were vigorously contested (Price, 1981; Porter, 1999; Shifrin, 2009). The health histories of the Turkish baths and associated hydrotherapies were symbolic of power struggles with emergent professional medical practice, where issues of regulation and ownership were informed as much by power and exclusion as by genuine medical concerns (Porter, 1999).

A second form of contestation for the sweating-places was a more spatial one whereby metaphors of healing waters had to compete with the then current traditions of spa and sea-bathing, though these too could be complementary as well as alternative (Durie, 2003a; Kelly, 2009). The role of folk medicine as a form of CAM, was played out in the naturotherapy setting of the sweat-house but also embedded through *dinnseanchas* and wider 'knowledgeable lay narratives' opposed to the more medicalized discourses of later hydrotherapy (O'Leary, 2000; Kavita, 2002; Cummins et al., 2007). Though applied to wider sauna identities,

Weir's comment can be effectively applied to relationship between sweating-cures, conventional medicine and wider psycho-cultural understandings of health; 'Before their secularization, saunas were part of the universal combination of religious, medicinal and psycho-therapeutic modes, which have only recently, like much else, been split off and compartmentalized by Western science and pseudo-science' (Weir, 1979, 190).

In their inhabited dimensions, the sweating cures reflected a range of ownerships. In their uniquely embodied curative form, where internal bodily waters were expelled in physical acts of detoxification and capillary exhalation, the sweating-cure was natural, personal and affective. In some of the more invasive douches and showers, Mackaman pointed out their role in providing a focused and directed embodiment of treatment onto particular body parts (Mackaman, 1998). While this can be seen positively as not so much a partial embodiment of health as a healthy embodiment of body part, alternative interpretations can be applied. In Foucauldian terms, it can signify wider subjugative medicalizations of the body from holistic subject to body-part object (Foucault, 1976).

From a cultural perspective, performances of health in place were fully visible in both the sweat-house and Turkish bath. At the sweat-houses, the rituals of enclosure, perspiration, wrapping and cooling were enacted in a simple and vernacular echo of the same rituals carried out in the Orientalist surrounds of the Turkish bath. While no formal 'script' existed, customary practice and a form of collective lay knowledge passed on the lay healing performance in and even through place. While different approaches existed, for example, wood was often substituted for turf in County Fermanagh, and the duration of the cure varied; the same broad performativities of self-landscape interaction took place (Richardson, 1939; Conradson, 2005a). In the Turkish bath, the performances of movement, rest and sweating reproduced global practices over 2000 years old (Jackson, 1990). In the material settings of both healing place, the rituals enhanced and to an extent completed, the experienced, enacted and felt dimensions of healing (Holloway, 2006; Rose and Wylie, 2006; Lorimer, 2008; Dunkley, 2009). An element of controlled performance also emerged at places like St. Ann's, where a large Belfry, sub-consciously evocative of its pseudo-Turkish muezzinic origins, was used as a call, not to worship, but to treatment and food (O'Leary, 2000). In this echo of the bell's discursive use in French watering-places as a form of managed temporality, issues of freedom, order and supervised practice were re-enacted and concretized (Mackaman, 1998).

In the mobile development of water-based treatments, shifts in medical culture and social mobility lay at the heart of therapeutic place experience. During and immediately after the First World War, St. Ann's transformation into a therapeutic space of rehabilitation and recovery for wounded and shell-shocked soldiers echoed the re-use of Craiglockhart Hydro in Scotland, recorded in Pat Barker's celebrated novel, *Regeneration* (Barker, 1996; Durie, 2006). In this reinvention as a rural healing place with a particular emphasis on mental health, it also reflected the wider medical world's tentative but necessary entry into that particular branch of health care (Philo and Parr, 2003). There was also a complex relationship developing through the eighteenth and nineteenth centuries around both health paradigms and access to health. As the hydrotherapeutic landscapes shifted from complementary and folk medicines to more professionalized and regulated systems of formal medicine, there were simultaneous shifts in utilization as the colonial and Anglo-Irish elites first gained, and later lost, power against a rising catholic middle-classes culminating in independence and a more democratic, though still morally supervised, society (Foster, 1988; Duffy, 2007). These shifts in socio-cultural worlds were also reflected in the place of the individual's negotiations of health in place, which were performed in embodied and mobile forms but were also shaped by wider controls and structures of a moral geography. For the poorer members of both rural and urban communities, the previous dependence on folk medicines and vernacular healing place, was giving way to more formal sites of public health care. Here a perceived improved provision was discursively sold through terms like modernism, progress and science, subtly implying that previous lay approaches to health were superstitious, ignorant and downright harmful. Yet in the next therapeutic landscape to be considered, sea-bathing, there were overlaps and contestations of these wider trends with a lingering and at times resistive attachment to informal medicine, lay understandings and self-medication as well as recursive metaphorical connections between health, bathing, liminality and leisure (Towner, 1996; Lenček and Bosker, 1998).

Chapter 5
Sea Bathing: The Thalassic Cure

Introduction: Tourism and Thalassotherapy

> It is well to note here, for those who are unable to make use of the open-sea bathing, that the powerful tonic agency of sea water is available, under cover, both hot and cold, at the establishment of Mr. Jones, of Martello Terrace. The town is well supplied with highly-qualified medical men, and some certificated nurses are available. Three dispensing establishments exist for the supply of practically every medical requisite that the invalid is likely to require. And should any special need arise, the metropolis can be reached in about half-an-hour by the frequent service of the Dublin, Wicklow & Wexford Railway Co. (Doran, 1903, cited in Horgan, 2002, 63)

In the above description of Bray in 1903, the town's reputation as a health resort is promoted with reference to the presence of formal medical infrastructures and the 'powerful tonic' of the sea-water in curative baths. In addition, proximity to the city is identified as a significant element in case of medical emergencies. Subsequent health narratives at the seaside have focused more on holistic and embodied aspects of exercise, fitness and more recently, on the potentially anti-therapeutic dangers of careless beach behaviours (Lenček and Bosker, 1998; Collins and Kearns, 2007). Yet older watering-place identities and the development of the sea-bathing resort as a therapeutic setting go back to the mid eighteenth century and it is from this period up to the 1930s that the chapter primarily focuses.

Though found in many cultures, narratives of sea-bathing as a healing activity had strong associations with England, where consumption of sea-water was first recorded at Scarborough in 1626 and filtered down to other countries, including Ireland, as part of a wider global trace (Walton, 1983; Corbin, 1995; Durie, 2003a). Healing activities included consumption of, and later immersion in sea-water, replicating and ultimately superceding the spa town in those practices and performances (Heuston, 1997; Urry, 2002). This increased fascination with the sea was driven by the same scientific and medicalized hydrological narratives that promoted health at the spa towns. Most often mentioned were Wittie (1667), Floyer (1701) and Dr. Russell's influential promotion of the sea-water at Brighton (1750), while an Académie de Bordeaux account in 1766 provided a similar boost in France (Walton, 1983; Corbin, 1994). Over time, 'bathing-in' supplanted the 'drinking-of' the water, and sea-resorts overtook inland spas in volume and fashionability, though some visitors still frequented both in successive seasons

(Shields, 1991; Kelly, 2009). While initial clienteles mirrored the spa town, being primarily the aristocracy, gentry, monied merchants and professionals, there were also narratives of early working-class uses of the beach (Walton, 1983; Corbin, 1995; Taylor, 2007). As at other therapeutic landscapes, social entertainments were devised to keep visitors in place and over time these supplanted the healing aspects, with the liminal pleasures at towns such as Brighton becoming more important than health as early as 1848 (Shields, 1991). The 'openess' of the sea-resort, especially in cultural, spatial and economic terms, featured prominently in critical narratives (Lenček and Bosker, 1998; Urry, 2002).

From a health perspective, the development of new thalassotherapies (literally cures from the sea), were part of an interest in cold water cures linked to wider hydrotherapy (Borsay, 2000; Hassan, 2003). In establishing a curative reputation for the cold sea water, cultural readings of the Romantic gaze were significant (Corbin, 1995). Until the beginning of the nineteenth century, the sea in Western cultures, reflecting contested health narratives, was seen as wild, forbidding and dangerous (Walton, 2000; Collins and Kearns, 2007). With the rise of the Romantic landscape tradition, people now turned their bodies towards rather than away from the sea and embraced it in a specifically experiential way (Wylie, 2007). Initial practices involved the consumption of sea-water, while bathing was of the short, sharp shock variety as evidenced by the formidable female dippers of eighteenth-century Brighton and Dun Laoghaire. Subsequently, practices similar to those used at spas, namely scientific verification of water source and curative power, were widely used (Walton, 1983; Bergoldt, 2006). Durie suggests that water-testing was not so important given that sea-waters, unlike mineral waters, were unvaried in their chemical makeup. The selling of place therefore became a stronger component at the sea-resort, with variations in the health product harder to justify (Durie, 2003a). While this view is contestable in environmental health terms, especially around pollution and water quality, it is helpful in emphasizing the importance of place to health at the seaside (Hassan, 2003). In addition, the length and openness of beaches provided greater spatial access to the curative product than in the narrow confines of the pump room (Shields, 1991; Towner, 1996). As the ingestion of sea-water went out of favour in the latter half of the eighteenth century, curative narratives, not yet focused on the value of swimming, shifted to more implicit elements of sea water, such as the chemical and restorative powers of fresh air, ozone, iodine and salt (Walton, 1983; Hassan, 2003). Gradually those health narratives developed in two directions, one related to hygiene and environmental health and the other related to exercise, sunshine and embodied aspects of fitness, beauty and health (Shields, 1991; Lenček and Bosker, 1998).

Similar histories were observable in Ireland. In particular there were accounts from holy well histories of a spiritual and curative link to the sea. Traditional beliefs in the healing powers of August's tidal waters, found across Western European coastlines, were repeated in Ireland with narratives of participation by both sick people and injured horses (Bord and Bord, 1975; Logan, 1981; Walton, 1983). The drinking of sea-water as a curative practice was observed on the coast north of Dublin as early as 1709 and along the coast of Clare in the 1740s, though its development as an upper class pursuit took place after 1750 (Beirne, 2006; Kelly, 2009). While many of the principal Irish seaside towns developed in the Victorian period and were opened up by railways, the use of coastal steamers in the early nineteenth century was just as important to towns like Kilkee and Portrush (Heuston, 1997; Byrne, 2005). However the arrival of the railways did bring, as elsewhere, a shift in social mobilities, whereby previously 'exclusive' resorts were invaded in large numbers by the hoi-polloi. In response to these social mobilities, resorts responded in different ways, some by attempting to manage, control and stratify the process, while others modified to cater for the new clienteles (Horgan, 2002; Davies, 2007). Individual resorts were also developed by the same sorts of entrepreneurial alliances found at the spa town, mirroring the mix of local and regional private investment and later, as in Britain, local government (Walton, 1983; Kelly, 2009).

Within Ireland, health histories of the seaside resort are surprisingly sparse though many tourism histories and geographies exist. While there are good individual resort accounts, particularly for Bray, Kilkee and Youghal, there are no general surveys like those of Walton in England or Durie in Scotland, though Kelly's work on the spa town usefully considers the relative role of sea-bathing at watering-places (Walton, 1983; Heuston, 1997; Durie, 2003b; Byrne, 2005; Davies, 2007; Kelly, 2009). The lack of material reflects some of the scalar and sourcist issues noted previously, though an emphasis on health histories in place, reflective of recent work in Devon, may stimulate a greater interest in the cultural, spatial and economic performances of health at the Irish seaside (Andrews and Kearns, 2005). In addition, there are a number of relatively unique thalassotherapy features such as seaweed baths which have a long vernacular tradition and a contemporary presence in Ireland which have yet to be fully researched.

In considering health identities and practices, resort's histories and spaces reflected wider social and cultural performances. As visiting the seaside shifted in identity from health to leisure and in clienteles from the 'upper' to the 'lower' classes, fashions and behaviours in places like Kilkee and Bray were mobile and mutable (Stallybrass and White, 1986; Urry, 2002). In particular the reputation of the seaside as a liminal place was enacted in social and moral contestations around embodied elements of dress and undress, modesty and morality and contestations between high and low cultures (Shields, 1991). This was reflected in how beach-fronts and the wider resort geographies were stratified and managed (Towner 1996, Walton, 2000). The seaside resort's organic attractions as places of restoration, retreat and fashion were subject to considerable economic modifications in an, at times, highly structured production of place (Davies, 2007). However, narratives of health, while often subsumed into narratives of leisure and pleasure, remained important in different forms at different times in different places (Hassan, 2003). Across time and space, the attractions of the beach and the sea, and the human utilization of those places as sites of healing and well-being were informed by human performances, which though connected to the social-cultural forces producing those places, also acted beyond them. Here Turner's notion of communitas and Bourdieu's idea of habitus were connected in communal tastes and habits that led to instinctive enactments in place; part of a long history of affective, therapeutic and embodied engagement between person, place and water. (Shields, 1991; Lenček and Bosker, 1998).

Around the Irish coastline, a large number of towns and villages had and have a sea-resort function, from large towns like Bray down to small villages like Ardmore (see Figure 5.1). Clearly sea-bathing can also be a private enactment, carried out in myriad coastal settings by individuals, but the focus here is on commercially developed and therapeutically-promoted landscapes (Heuston, 1997; Duffy, 2007). The boxes in this chapter relate to three towns; Bray (Co. Wicklow, see Box 5.1) which as noted, acted as a primary resort for Dublin, while Kilkee (Co. Clare, see Box 5.2) and Portrush (Co. Derry see Box 5.3) were popular resorts for residents of Limerick and Belfast respectively. Other resorts chosen for their relatively long histories include the southern resorts of Youghal (Co. Cork), Tramore (Co. Waterford) and Ballybunion (Co. Kerry). The final resorts include Lahinch (Co. Clare), Dun Laoghaire (Co. Dublin) and the County Sligo villages of Strandhill and Enniscrone. The latter two were chosen in particular because of the presence of commercial seaweed baths.

Embodied Performances of Health: Seaweed and Submersion

Stories of the health benefits of sea-water were often vague, yet despite regular biomedical contestation, they sustained an embodied therapeutic reputation from the mid seventeenth century to the present-day. 'There is a physick in the sea' was the simple belief, linked to earlier magico-spiritual narratives, of working class visitors in Lancashire in the eighteenth century (Walton, 1983). Although the drinking of and bathing in, sea-water was at the forefront of early performances of health, these began to retreat by the early decades of the nineteenth century, as the social practices of leisure and entertainment made ground. In Britain, this threat to the original health identities of place meant that many of the resorts began to update those narratives, though still with embodied and experiential elements (Hassan, 2003). This shift was particularly relevant from a therapeutic landscape perspective, as it focused closely on the benefits of sea-air, shaped by local geographies of micro-climate, topography, soil, prospect and aspect (Borsay, 2000; Durie, 2003b, Hassan, 2003). This discursive selling of health and place from the mid nineteenth century also tallied with the modernist rational demands for statistical evidence, which in turn was used in competitive health promotional activities between resorts (Hassan, 2003). Within Ireland this was reflected in the suggested benefits of the views and sea-air at Bray and Tramore (Horgan, 2002). Knox's account of Kilkee noted:

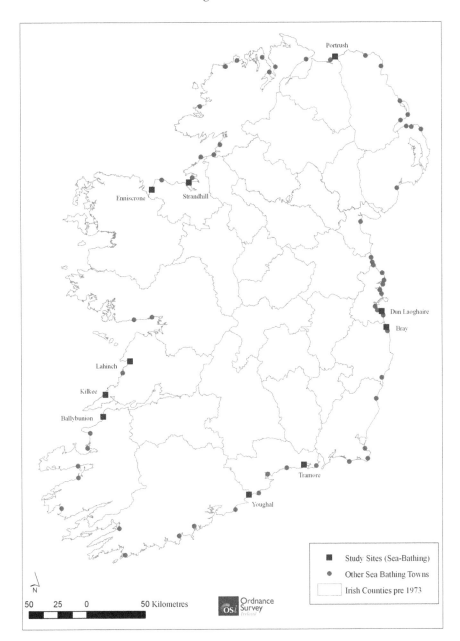

Figure 5.1 Map of the Distributions of Sea Resorts with 10 place-study locations

Source: Boundary Maps, Ordnance Survey Ireland. © Ordnance Survey Ireland/Government of Ireland Copyright Permit No. MP 008009.

It affords excellent facilities for bathing. Persons labouring under spinal disease enjoy here the benefits of the plunge bath, being carefully attached to the beach by boards adapted to the purpose and immersed with all due gentleness in the water. The ... cliffs on both sides of the town afford delightful walks, where the bracing sea air, charged with imperceptible saline particles, may be breathed in all its freshness and purity. (Knox, 1845, 313-314)

These notions of freshness and purity used in the promotion of the seaside, contrasted with the filth of the city and reflected colonial Victorian obsessions with cleanliness as both a hygienic reading of health in place and a modernist concern with order and civilization (McClintock, 1995; Kearns, 2007). While this discourse was less commonly used in rural Ireland than in urban Britain, the relative size of Dublin and Limerick meant that narratives of a temporary flight from the diseased city were associated with Bray and Kilkee (Beirne, 2004). In Kilkee, Knox's account also notes the cleanliness of the water when compared to more polluted British examples (Knox, 1845).

Initial performances of the sea-water cure, were of the ingestive and immersive kind (Walton, 1983; Durie, 2003b). As with the consumption of mineral waters at the spa, the effects were purgative and unpleasant with dosages spread out across the day. The presence of a couple who ran a lodgings house and provided attendance for 'those who have a mind to drink salt-water' was recorded at Portrush as early as 1761 (Horgan, 2002). By the later decades of the eighteenth century practices had moved towards bathing as the preferred form of healing activity. The role of 'dippers', usually women of formidable strength and character reminiscent of earlier watering places, was imported in the late eighteenth century from England and were found in Kilkee, Dun Laoghaire and Bray. Typically, as most bathers of the period did not swim, they would be towed out in bathing boxes, for both modesty and safety and, upon their emergence at sea level, would be vigorously dunked several times in the sea (Shields, 1991; Towner, 1996). Such a practice was carried on by the 'washer-women' of Dun Laoghaire who complained about the new coastal railway in 1834, which in restricting access to the beach, threatened their 'dunking' practices of over 50 years (Agnew, 2003). In time, bathing unsupervised became the norm, though the widespread ability to swim was a much later phenomenon (Lenček and Bosker, 1998). The customary practice of swimming was never a part of vernacular coastal cultures, especially in the west, where fishing communities saw the sea in fearful, fateful and by extension health-risk terms (Wood-Martin, 1892; Heuston, 1997).

Box 5.1 Bray, County Wicklow

Bray's creation as a sea-resort was essentially copied from other towns in the British Isles. Originally a small fishing village on the coast south of Dublin, that proximity made it a prime target for development, aided by local entrepreneurship. By 1845 its identity as sea-bathing town was already evident, with baths, promenade and accommodation, a process intensified with the arrival of the railway in 1854. This opened Bray up for the day-trippers as well as the elitist longer-term summer visitor, which brought tensions around social stratification. The period between 1855 and 1863 was a particularly intense one, with three new hotels, nine separate terraces, and three churches built in this period, along with the short-lived Turkish Baths (Davies, 2007). While the Turkish Baths and the separate Bray Hydropathic Establishment, were short lived, broader health identities were exemplified in the construction of hot, cold and sea-water baths at Martello Terrace in 1861. Early marketing played heavily on Bray's role as a health resort and some guidebooks even included comparative tables emphasizing Bray's low death rates (Flinn, 1888). That health identity, linked to narratives of sea-air and sea-water remained strong into the 1920s and 1930s. Spaces of health included the sea, baths and ancillary elements such as accommodations, amusements and open spaces. While large hotels were a feature of the town, the southern end contained older lodgings house which marked a stubbornly local presence. Another important element were bathing boxes, provided at Bray by private individuals but also the railway company and the local Loreto Nuns.

Figure 5.2 Swimming Baths, Bray, Co. Wicklow, c1930
Source: Image from the photographic archive of Irish Heritage Giftware, info@ihpc.ie.

The material ownership followed standard colonial models, being originally developed by private owners and investors while Bray Town Council became more involved after 1870. Elaborate plans were made via the Bray Pavilion Company to make the town more closely fit a British model, by building Assembly rooms and a pier, but only a Ladies Baths were actually completed. While the railway was significant in opening up the town, new transport forms such as the bicycle and motor car gained importance after 1880. The marketing of Bray was linked to a range of discursive tactics such as the spatial appropriation of the nearby Wicklow Mountains and also an invocation of Brighton in particular. Bray Town Council were also involved in the social regulation of the town and the seafront, in line with then current mores, and the spatial segregations of gender and class were important in shaping the inhabitation of the town. While the wealthier clients frequented the hotels and terraces to the north, the smaller and cheaper lodgings houses, were frequented by poorer visitors. From a gender perspective, and in keeping with similar divisions in other resorts at this time, ladies and gentlemen's bathing areas were separated and strict by-laws introduced to control any form of mixing. While relatively successful prior to World War I, such segregations broke down after the 1920s, a period when narratives of health at the resort began to disappear. In time, more liminal social discourses of escape, otherness and the carnivalesque began to shape Bray's identity.

In its embodied aspects, sea-bathing represented some important psycho-cultural understandings of healing and well-being associated with water. All through history, immersion in water, as well as being a traditional curative act, was enacted as a spiritual ritual of initiation and rebirth in many world religions (Sellner, 2004). In a more secular sense, swimming and bathing were also embodied acts with contestable health meanings (Lenček and Bosker, 1998). As enactments they were tactile, affective and often involved the whole body in a near naked contact between self and therapeutic element (Smith, 2005). Historically, the new-found pleasure for the patient of a swim and exercise in water, while being excused the sometime penance of actually drinking it, was considered a 'let-off' by early Victorian bathers. Here the ability to mix pleasure and therapy and enjoy a sugared pill was duly noted (Walton, 1983). In the subsequent submersive acts of swimming, diving and floating, whether in the sea-water or at a nearby lido, physical activities also benefited health and general fitness (Hassan, 2003). In bathing, either in the sea itself or at sea-water filled baths in Bray, Dun Laoghaire, Portrush and Kilkee, the implied benefits of the sea-salts, were regularly used in place promotion such as Hogan's lengthy medicalized promotion of the value of baths at Kilkee (Hogan, 1842; Heuston, 1997; Davies, 2007). The sea-baths also provided a safe-haven (as at Dun Laoghaire) for those who were afraid of the sea; 'Safe bathing is available for those who dislike the open sea, there being two swimming pools – one for ladies and one for gentlemen – and these are replenished regularly with fresh sea water. A children's pool is also provided' (Thom and Sons, 1911, 67). Subsequent discourses of health and embodiment from the 1920s on further emphasized fitness and even beauty aspects of the 'sea cure'. While heliotherapy and tanning were, not surprisingly, rare in Ireland, the mixed virtues of reliable fresh-air, unreliable sunshine and a culturally-understood discourse of generalized health benefits continued to draw people in their thousands to Portrush in the 1940s as much as it did to Kilkee in the 1840s (Heuston, 1997; Byrne, 2005).

One bodily performance of health that was and still is relatively unique to Ireland was the widespread use of seaweed baths. In the early decades of the twentieth century, there were allegedly over 300 such baths dotted around the Irish coasts (Voya Seaweed Baths, 2009). There were seven in Strandhill alone, and some lodgings houses sold a package of accommodation, ale and a seaweed-bath (Voya Seaweed Baths, 2009). As a natural product of the sea, and plentiful in the West of Ireland, seaweed had long been used as an agricultural fertilizer but also as a healing product and even a form of food (Guiry, 2006). This interest in the healing properties of certain seaweed species, especially wracks such as *Fucus serrata* and *Fucus vesiculosus*, has been sustained in contemporary seaweed baths in Strandhill, Enniscrone, Ballybunion, Kilkee and Newcastle (Finlan, 1993). In terms of the specific 'cures' associated with the use of sea-water, these were mostly linked to skin conditions and arthritis. In the case of the latter, Daly's Seaweed Baths at Ballybunion (now closed) noted that; '... the natural goodness of bathing in salt water is considered to bring welcome relief to general muscular stiffness, sports strains, stress, tension, arthritic and rheumatic pain'. This relaxant was augmented by seaweed's iodine content, which acted as a natural antiseptic. This use of seaweed for cuts and wounds reflected global watering-places practice such as an equivalent use by the Māori in New Zealand of hot springs (Rockel, 1986). At Collins' Hot Seaweed Baths in Ballybunion, more biomedical benefits are listed in leaflets stating that seaweed has a detoxification and relief function for a range of skin and respiratory conditions. The proprietors of the baths also recount narratives of successful cures, especially in a case of eczema, where a series of daily visits across a fortnight's holidays, proved effective in managing and reducing the symptoms. It is in these aspects of health management and the alleviation and reduction of chronic diseases and ailments that many of the therapeutic landscapes discussed in the book are at their most effective. Along with the traditional seaweed baths still operating at Ballybunion (Collins-Mulvihill's), Enniscrone (Kilcullen's, see Figure 5.3) and Strandhill (Voya-Walton's), there are newer seaweed baths in Newcastle, Kilkee, Bundoran and several other towns. Seaweed baths were historically recorded at Kilkee, Portrush and Dun Laoghaire, while other types used for medical and recuperative purposes included Russian, sulphur and alkaline baths at Dun Laoghaire's Victoria Baths. Russian or vapour and iodine baths were also run by Hugh Hogan at Kilkee (Hogan, 1842).

As in other therapeutic landscapes, associations between improved health and the promotion of the sea-resort were contested. For adherents of the curative mineral spa, the absence of verifiable and reliable indicators such as specific minerals, meant sea-water was never considered as effective (Durie, 2003a; Beirne, 2006). As with other hydrotherapies discussed in the book, the presence of informal practitioners and practices were also noted with concern (Porter, 1989;

Hembry, 1997). Children were fairly brutally dipped at Kilkee while a lady was observed at Bray literally walking the plank on a narrow pier while being held on a rope by an old lady (Heuston, 1997; Davies, 2007). Given new public health research on the relative dangers encountered at the seaside, particularly drowning, it might be expected that such accounts, of the former at least, would be found in Ireland (Collins and Kearns, 2007). Such narratives were occasionally found in discussions of the Irish seaside before the 1940s, though it should be noted that mass recreational swimming only developed after this date (Lenček and Bosker, 1998). Accounts of regular drownings are recorded at Kilkee, including a young couple drowned in the late 1880s. The 'precarious pleasures' of bathing were noted in the 1860s in relation to the unfortunate drowning of three women at Portrush, who in the absence of suitable bathing-boxes, were washed off some secluded rocks (Irish Builder 1861). Here improved safety was linked to a call for more risk-averse behaviours while also noticing the possible fear of immodesty as a causative factor. Other contested understandings of health in place related to water quality and the risks of pollution. While smaller Irish resorts were less susceptible to unclean waters than the larger British or French resorts, Bray Town Council were sufficiently alarmed as to draw up plans for improved sanitation in the town in the 1880s to protect public health (Clare, 1998).

Figure 5.3 Kilcullen's Seaweed Baths, Enniscrone, Co. Sligo, 2009

At the sea-resort, the varied powers of the healing waters were reflected in shifts from embodied and ingestive performances to later immersive, even aerative enactments. While it is tempting to see this shift in embodied terms as a move from the physical to the more imaginative, the body still lay at the heart of the sea resort cure. Bathing played a continuing role, but in place terms, being near and around the therapeutic power of the sea was often as important as being in it. In the wide expanses of Tramore or Lahinch beaches, there was literally space for everyone. In terms of the self-landscape experience, the narratives of danger and safety at Portrush and Dun Laoghaire were mirrored at Kilkee where shelter and protection from the wild Atlantic waves continued to be part of the selling of place. Finally the relative importance of the health experience at the seaside were threatened and ultimately surpassed by a more social performance, though earlier practices of health still lingered, not just at the seaweed baths of Enniscrone or Strandhill, but also on the beaches themselves.

Cultural Performances of Health: Liminality and Restoration

In a manner very similar to the spa town, a range of cultural performances of health shaped place and behaviour at the sea-bathing resort (Urry, 2002). The social structures of the watering-place, in their clienteles, enactments and circulations, were well established from the spa and it was often simply a case of transferring those forms to the new sea-settings. Brighton for example, was often described as 'Bath-by-the-sea' and indeed the upper classes moved between the two in successive seasons, thereby providing a health/place legitimization of both (Farrant, 1987; Shields, 1991; Hassan, 2003). Yet even from the 1860s the number of direct narratives of health at the seaside had already begun to decline in Britain, replaced by narratives of leisure, which while primarily social, were also seen, in their restorative and restful aspects, as providing a form of health justification to that narrative. Hence, although by the 1880s there were still a few stories of miraculous sea water 'cures' for visitors, the nature of therapeutic practice gradually changed from medical and curative to the restorative and health maintenance features of holidays in good summer weather (Shields, 1991).

This pattern was repeated in Ireland, especially as the cultural values of the dominant Anglo-Irish were enacted in colonial social performances of health. At Bray and Kilkee, the initial health narratives were replaced by a new focus on the beach as a focal point for entertainments and public utilization. More broadly, the spaces of cultural performance at the seaside, especially in the long promenades at almost all of the studied places, certainly extended the more constrained sites of the spa town. Lenček and Bosker recorded that the beach was a site of both social spectacle and social segregation acted out in place and this was certainly true of performances amongst the elite users of the seafronts of Portrush, Bray, Tramore and Kilkee (Lenček and Bosker, 1998; Davies, 2007). As the resorts became more socially mixed, a spatial segregation emerged at Bray, though this was a self-selective rather than enforced practice such as at Bridlington, where tolls were introduced to keep up the social tone (Walton, 1983). More importantly the health and leisure aspects of the seaside resort were, for many, enacted in precisely restful and restorative performances of gazing at the sea and simply lounging about; practices still central to beach behaviours. These were noted in a particularly hot late summer in Kilkee in 1882, with a 'Strand ... draped in feminine forms' (Marrinan, 1982). Here 'landscape-as-being' was enacted in restful, still and static ways while the multi-sensual and arguably affective therapeutic enjoyment of nature was reflected in the sights, sounds and smells of water, rocks and birds (Conradson, 2005b and 2007; Lorimer, 2008). By contrast, there was a parallel affective identity associated with the noise of the constructed sea-resort, a carnivalesque and at times cacophonous soundscape which also formed and forms the life of a place (Gibson and Connell, 2005). In that same summer in Kilkee, rowdy sports days were held in front of Moore's Hotel, while musicians played and 'shawl' dances, competitions in which the winners won shawls, were held on the strand.

At the seaside, there were both typical and unique cultural performances. Irish resorts, in a process of colonial and cultural imitation, reproduced practices found at the British resorts. Bandstands and promenades were built at Bray, Youghal, Portrush and Kilkee, and social performances of strolling, assembly, listening to music and polite conversation formed an important part of the daily routine. In these acts of assembly and conversation the formations of basic cultural identities were reinforced by the company of like-minded and like-behaved people (Bourdieu, 1977; Breathnach, 2004). Daily activities at Kilkee, acknowledged in the mid nineteenth century as a solidly Protestant location, involved walking, bathing, rock-pool exploration and donkey-rides (Heuston, 1997). The latter activity, recorded by Thackeray in Kilkee in 1842, was a surprisingly typical (and still current) activity at beaches across Western Europe though its origins are hard to place. Yet at Kilkee, the unique aspects of the setting were also absorbed into social practice, in the trips, often in open *currachs*, to visit sea-caves (de Bovet, 1891). Open sailboats, know locally as 'drontheims', were also used for excursions at Portrush to the offshore Skerries islands (Young, 2002). Sports were also visible in other forms through the regattas at Bray, Portrush and river cruises out of Youghal up the River Blackwater, while sailing remains at the heart of Dun Laoghaire's identity up to the present day (Orme, 1966; Young, 2002; Davies, 2007). The presence at many of the resorts of extensive sand dunes, the raw material for the classic 'links' golf course, was ideal for the development of the sport as a leisure activity and element of colonial identity (Walton, 1983). At Ballybunion, Portrush and Lahinch these natural features could literally be landscaped into therapeutic sites of leisure. In addition, several of the resorts including, Youghal, Tramore and Kilkee had race-tracks on or near the beach, where further excuses for recreation and of course, gambling were to be had.

Box 5.2 Kilkee, County Clare

One of the earliest functioning sea-resorts in Ireland, despite its remote location, sea-bathing was recorded as a curative performance at Kilkee in the 1780s and it was an established resort by 1820. An initial identity as family resort was already in place with access by steamer from Limerick city via Kilrush. This water-based opening up of access was unusual for Ireland though it mirrored practices in England and France (Corbin, 1994). Noted by Knox as the most fashionable resort on the west coast of Ireland in 1845, the facilities in the town were aided by the construction of new terraces and hotels, though most houses in the village were pressed into service as summer lodgings. Entrepreneurship was provided by the local landowners and hoteliers meeting a demand for a seasonal population of up to 4,000 in the mid nineteenth century. Kilkee was probably at its height as a sea-resort by the 1860s. In the first decades of the new Irish Free State in the 1920s and 1930s, Kilkee, in common with most sea-bathing resorts, struggled. The traditional protestant Anglo-Irish clienteles were much reduced post-independence and gradually new transport forms such as motor-buses helped a new Catholic professional middle-class revive Kilkee as a resort. In the 1920s an attempt was made to remarket the town as the 'Biarritz of Ireland' which brought with it electrification.

Figure 5.4 Strand at Kilkee, Co. Clare, c1890
Source: Courtesy of the National Library of Ireland.

From its foundations, performances of health had an important role in the production of place. Kilkee's reputation as a sea-bathing, and by discursive association, health resort was based initially on the consumption of and bathing in sea-waters in the town's protected bay and on the Duggerna Rocks. Knox noted the presence of cold, hot, sea and seaweed-baths in 1845 but in a hint at a developing demand in hydrotherapies, noted the need for douches and forcible steam bath treatments. Discourses of health were also supported by reference to several nearby mineral spas which added to the town's curative reputation while immobile patients were carried down to the water on boards to immerse themselves in the sea-water (Heuston, 1997). In addition to the water, wider health discourses, specifically focused on the value of ozone and sea-air for respiratory and other conditions, were also employed. More widely the therapeutic potential of place was invoked through the surrounding landscapes, which were seen as inspiring to the spirit, while a notional peacefulness and quietness was also used in the commodification of the town as healthy place. In its linked social inhabitations, the town may not have been as quiet as the sellers of health might have had one believe. As with most watering places, especially one where the visitors were on holiday both physically and metaphorically from their normal lives, a number of liminal behaviours were enacted in place including drinking, gambling and match-making. Over the years the performances of health in place declined and were expressed through more restorative and restful dimensions of well-being.

While many of the accounts of the sea-bathing resort focused on the curative and cultural practices of the social elites, an earlier working-class culture of sea-use was found in Irish resorts (Walton, 1983; Corbin, 1994; Towner, 1996). The British pattern of wholescale shifts of urban populations for 'worker's week holidays, was not common, though Portrush did experience regular influxes of visitors from Scotland and Bray received growing numbers of tourists from England and Scotland in the early twentieth century (Byrne, 2005; Davies, 2007). However early European littoral practices of visits to the sea-side by the rural working-classes were enacted in places as diverse as Lancashire, the Basque Country and Kilkee, especially by farming communities at the end of harvest (Walton, 1983; Urry, 2002). At Ballybunion, even into the 1940s, the tradition was for farmers and their wives to come for a ritual end-of-harvest dip in the sea, often in the evening or at night, depending on the tide, where they would bathe in their nightdresses and drawers. Similar groups of rural dwellers from Kilkenny, known locally as 'dournawns', were recorded at Tramore in the nineteenth century (Taylor, 1990). Of particular interest was the recording of a practice enacted around Sligo. Here, poor rural dwellers in the mid-nineteenth century, known as 'Sea-Pikes', would go the seaside en masse for several weeks holidays, a common enactment seen in specifically curative terms, but undoubtedly with social dimensions as well:

> I was surprised to meet every few hundred yards on the road (i.e. from Bole to Sligo) carts heavily laden with country people, many of them of the lowest order, and with different articles of furniture piled up, or attached to the carts;

and I learned with some astonishment that all these individuals were on their way to sea-bathing. This is a universal practice over these parts of Ireland. A few weeks passed at the seaside is looked upon to be absolutely necessary for the preservation of health; and persons of all classes migrate thither with their families. On my way to Boyle, I met upwards of twenty carts laden with women, children and boys. One may ask how the people afford this annual expense? But there are numerous cabins and cottages at the lower end of Sligo, on the Bay, in which a room is hired at 1s. 6d. per week; this is almost the whole expense, for all carry with them – besides their beds and an iron pot – a quantity of meal, some stacks of potatoes and even turf is there is room for it. (Inglis, 1834, quoted in Wood-Martin, 1892, 410)

Quite apart from the barely suppressed astonishment that the peasantry might aspire to take a holiday, the quote is especially useful in establishing local customary practices of sea-bathing with well-understood and vernacular links to 'preservative' health. Here assumptions of sea-bathing as elite performances of health are contested by more vernacular inhabitations.

Contestation was also to be found in the extent to which discourses of health could be sustained in places which had clearly shifted to more leisure and pleasure focused identities (Hassan, 2003). As at the spa, the specific biomedical benefits of sea-water were more commonly lionized than dismissed. Bathing in the sea came in for much less contestation than earlier hydrotherapeutic forms. In the transformation of the sea-resort into a site of leisure by the mid-nineteenth century, narratives of health were not expressed in a medicalized sense, and therefore, it could be argued, presented no threat to the rapidly establishing medical profession. While medical staff did exist at the sea-resort, medicalized accounts were rare and the absence of the specific hydropathic facilities found elsewhere was noted at Kilkee: 'There is every facility for cold, warm. Salt water and shower baths at Kilkee, as well as the sea-weed baths, administered there in great perfection; but the addition of an effective apparatus for the douche or forcible steam bath, is a desiridatum' (Knox, 1845, 318).

Yet, there were 'other', both literal and figurative, identities and enactments taking place at the sea-side resort which reflected alternative identities enacted at holy wells and spas. A number of cultural historians have written at length on the role of liminality and the establishment of transgressive identities and behaviours in marginal places, of which the seaside resort was a particularly good example (Stallybrass and White, 1986; Lenček and Bosker, 1998). At places like Brighton and Coney Island, the seaside was seen as acting as a type of social safety-valve where liminal and carnivalesque activities and performances were permitted (Bakhtin, 1984; Shields, 1991). In addition, the geographical position of the beach, nominally a limen, a 'free' inter-spatial and inter-tidal zone, reflected earlier meanings and identities of chthonic and human landscapes and encouraged identity inversion and freedom (Stallybrass and White, 1986). At Kilkee, the social openness was reflected in its reputation, like Lisdoonvarna, as a marriage market, an identity still present in 1944. Around 100 years earlier, the evening entertainments at the resort were noted in a letter to a Limerick newspaper; 'At night the real divarhsin' comes on. Everything imaginable gattin forward. Lashing o'punch, cards, pitch an'toss – dice or a devilment. God help the bathing boxes about two in the morning' (Heuston, 1997, 18).

Informal practices of drunkenness and gambling were identifiable here, though it is not entirely clear what was going on in the bathing boxes. In the later Victorian period, the sea-resorts were also sites of class contestation, wherein the original Protestant elites, as at Bray and Tramore, came under threat from the new and poorer classes who had begun to flock to the seaside. Here as elsewhere, these new 'arrivistes', often day-trippers, were seen in pejorative terms, for their uncouth behaviours and immoral habits. Particular concern was expressed in Bray in 1855 at the proposal by the Mayor of Dublin for former attendees at the infamously carnivalesque Donnybrook Fair (ultimately banned for its drunkenness, immorality and violence) to spend that holiday in Bray instead (Davies, 2007). As was often the case with spatial segregations, when one group's exclusive space was invaded one simply moved elsewhere. The higher-class holiday-makers subsequently moved down to coast from Bray to Greystones, which was known in 1901 as Rathmines-Super-Mare, the latter being a particularly affluent Dublin city district (Davies, 2007).

Overall, the communal cultural performances of the sea-bathing resort were reflected in their discursive links to health and leisure and to native and colonial practices in space. Liminal performances in place could also be seen in native and Colonial terms. In its essence the colonial project was built on narratives of order, progress and civilization and these were reflected in the development of public spaces as on Bray and Portrush seafronts and in the planned promenades, bandstands and bathing boxes (Sidaway, 2000; Kavita, 2002; Davies; 2007). There was an expectation of equally ordered and correct behaviours in these spaces (Shields, 1991). While this was understood by the original upper-class visitors to Bray, Portrush and Tramore, that understanding was both threatened and altered by more working-class and native behaviours. Trips to the seaside were much more communal than to the spa or hydro and had surprisingly strong vernacular histories as well as being reflective of the later habitus of the Anglo-Irish ruling classes. Initially curative and health-focused in their intention, the performance of the visit shifted from health to leisure pilgrimage. Yet the health narrative of the sea-side never completely disappeared and in their metaphorical associations with retreat, relaxation and rest, they mark a valuable link to more contemporary narratives of de-stressing, restoration and as notional spaces of respite (Conradson, 2005a; Smith and Kelly, 2006).

Spatial Performances of Health: Carmen and Ozone

The Irish seaside resorts reflected the significance of the spatial within wider performances of health. The production, or more specifically, reproduction of place was an important element, reminiscent of the replication of other globalized curative forms (Corbin, 1994; Hoyez, 2007a). In such settings, imitation was a deeply sincere form of flattery, even if one should remain mindful of Ireland's cultural and geographical position as both imperial subject and imperial object (Kavita 2006). One distinct advantage of Irish resorts were their long sandy beaches. These were common to all coasts and were relatively unspoiled and uncrowded, especially when compared to the hyper-development of the South Coast of England, the Riviera or the Jersey Shore (Towner, 1996; Lenček and Bosker, 1998). With the exception of the groynes at Youghal and the promenade at Bray, very few attempts were made to shape the shoreline as an enactment of cultural power over nature (Wylie, 2007). The resultant naturalness and indeed stillness and quiet, remain strong aspect of both the social and therapeutic appeal of the seaside in Ireland. The classical elements of the sea-resort ensemble, such as hotels, lodgings, bathing-boxes and baths, as well as fairgrounds, amusement arcades and donkey-rides were also evident. The classic British seaside pier never really featured in Ireland, despite putative attempts to build one in Bray in the 1870s (Walton, 2000; Davies, 2007).

Some of the invocations of health benefits to be enjoyed at the sea incorporated the presence of the chemical ozone. In Britain, ozone was regularly used in the therapeutic promotion of sea-air, even before its chemical composition was firmly established as O_3 in the 1860s. At Tramore the chemical and its attributed benefits were included in a rather overblown association of health and place that:

> ... lures the society pilgrim in search of health and pleasure. Here Tramore steps
> in and fills our right hand with a bouquet of refreshing offerings for the tourist;
> a strand carpety and expansive; an Atlantic breeze laden with ozone, a bathing
> resort to suit all tastes and abilities; possessing rail accommodation to the nearest
> port of the most efficient class, with a sea coast broken, varied and abundantly
> picturesque. (Egan, 1894, 605)

Here the beach, surroundings, transports and resort facilities were used to combine with the curative in a complex narrative with a specified intention of attracting the evocatively named 'society pilgrim', to spend time at a beach. While the air at different sea-resorts was confusingly described as being both soothing and bracing, the suggested effects were the same. In deriving its air, ozone, iodine and other curative characteristics specifically from its proximity to water, the sea cure could be seen, in contrast to the previous chapter, as a sort of 'salt-air bath'. Hassan noted that;

The sea was conceived as a gigantic receptacle of mineral water, which impregnated coastal air. Thereby it was transformed into an elixir of life. Many resort guides claimed special health-giving attributes for the prevailing winds, which were supercharged with oxygen or ozone, having passed over the ocean or vast moors. (Hassan, 2003, 76)

While the presence of both native and colonial utilizations of the sea for healing purposes dated back to the early-mid-eighteenth century, the full development of the resorts did not take place until the early decades of the nineteenth century. As in Britain, the horse and cart and later the steamer, opened up access to resorts like Kilkee and Portrush (Heuston, 1997; Byrne, 2005). Prior to the arrival of the railway in the late 1830s, the quality of roads was a major drawback but the building of more reliable turnpikes in the early nineteenth century meant that access from cities to local resorts was improved at Tramore and Bray. More importantly water itself was used via the introduction of a regular and fast steamer service through the Shannon estuary bringing Limerick's leisured classes to Kilkee from the 1820s, some 70 years before the railway (Heuston, 1997). Knott's 1836 account of the hustle-and-bustle at Kilrush, where passengers alighted for the final leg of the trip, attested to the established business by this time (Knott, 1997). The sea was even involved in an international trade, with twice weekly ocean-steamer links between Portrush and Glasgow, Heysham and Liverpool in the 1880s (Boyd, 1891). Here the links were as much diasporic as colonial, given the post famine migrations to those settings (Horgan, 2002).

While such therapeutic economies did exist prior to the arrival of the railways, the latter were crucial in subsequent waves of resort development. Indeed the world's first commercially developed passenger train line was built to connect Dublin city with its maritime suburb, Dun Laoghaire, where sea-bathing had been practiced since the 1780s (Quinn, 2003; Duffy, 2007). In time other city-satellite resort catchments developed such as Belfast-Portrush, Cork-Youghal and Waterford-Tramore, while even the relatively remote resorts of Lahinch and Ballybunion were joined to the railway network by the end of the century. Alongside the passenger services at Dun Laoghaire, other world firsts included the Lartigue Railway (1888–1924) a commercial monorail between Listowel and Ballybunion and the electric tramway between Portrush and Bushmills (Young, 2002). Finally the role of the jaunting car, an essential connecting service between transport terminal and holiday home, was an important element. Ironically, given the identities of 'car-men' as the antecedents of the taxi-driver, narratives of greed and over-charging were commonly found. The rude behaviours and outrageous fares demanded by the 'jehus' of Kilrush to bring visitors to Kilkee were mirrored in similar complaints about the Carmen of Bray (Heuston, 1997; Davies, 2007).

There were also aspects of exclusion and exclusivity which were affected by the opening of access (Towner, 1996; Walton, 2000). As was the case in mainland Britain, the exclusivity of resorts was altered by new clienteles, more humble in origin and yet more numerous in volume (Walton, 1983; Hassan, 2003). In

Ireland, the gradual process of de-colonization, already under way by the end of the nineteenth century, and accelerated after independence, was reflected in a democratization at the beach as at the spa and hydro (Breathnach, 2004). In a resistive narrative at Kilkee in the mid-nineteenth century, the local nickname for the Anglo-Irish was *ruachach* (literally, 'red shits'); linked to their iron-rich meat diets which the locals could not afford, and worse still, had to collect (Heuston, 1997). Later visitor lists at Kilkee in 1908 still identified over 70% as Protestant, though there were subtle hints that people 'lower down the social scale' were holidaying there (Heuston, 1997). Though the Catholic middle-classes began to appear at Kilkee, Tramore and other resorts at the turn of the century, the lack of paid holidays, which did not appear in Ireland fully until the 1930s, still restricted native clienteles to the more affluent professional classes of teachers, *gardaí* (policemen) and nurses; while the appearance of clergymen (protestant, catholic and non-conformist) mirrored clienteles in the other watering-places. Finally the geography of the clienteles reflected proximities to large population centres and wider regional affinities. Thus Kilkee was also known as the place where Limerick people went (and still go) while Youghal became the resort of choice for the working classes of Cork after World War 2 (Hackett, 1994; Heuston, 1997). As well as being the main resort for Waterford, Tramore was also a traditional venue for visitors from neighbouring inland counties such as Kilkenny and Laois, reflecting regional patterns of origin-demand and customary communal affinities of place common in England and beyond (Walton, 1983).

At the resorts, the traditional clienteles and catchments, especially in the mid nineteenth-century, were primarily colonial, though within a century, these were succeeded by more native inhabitations. Leisure and class were linked as the Anglo-Irish elites were the only ones who could afford the money and the time to access the sea-cure. This reflected patterns noted at the spa and hydro, given that the stay, rarely for less than a couple of months, was a considerable financial and temporal investment. Typically, wife and children would take up attendance at the resort, while the paterfamilias would either come down at the weekend, if close to the home (such as at Bray, Tramore or Youghal) or for a week's holiday (as at Kilkee or Portrush). These inhabitations were more mixed at a resort like Bray, especially given its accessibility to Dublin, and 4,000 day-trippers and seasonal guests were counted there on a Sunday in 1861 (Davies, 2007). At Kilkee, in 1845, the same number of visitors were counted during the season, which expressed how relatively well-established a resort it was, despite its remote location (Knox, 1845; Heuston, 1997). In addition, other representatives of colonial power, namely the Army, were also to play a role. The founding of Lahinch Golf Club, an iconic element in the town's historic and contemporary identities, was 'discovered' by members of the Black Watch Regiment based in Limerick and developed to serve their recreational needs (Clare County Library, 2009).

Box 5.3 Portrush, County Antrim

One of the oldest towns in the country, Portrush was also among the first to record a healing place identity with both sea-bathing and sea-water consumption recorded in 1761. Knox (1845) recorded the town as one of the most frequented sea-bathing places in Northern Ireland with a good hotel, baths and the presence of steamers passing along a route connecting the coast and Glasgow and Liverpool. By the 1880s local accommodation and accessibility were improved via the development of a large hotel, the Northern Counties and an associated railway line. More broadly, Portrush benefited from its proximity to the Causeway Coast and its most famous feature, the Giant's Causeway, for which Portrush became the effective gateway. Social facilities were developed via the construction of a new golf course in 1888. A range of terraces were built to provide accommodation to a developing summer population, mostly developed on the promontory separating the town's East and West beaches. As at other resorts in Ireland, Portrush was at various times referred to as the 'Brighton' and even 'Biarritz of Ireland', a discursive association commonly used to create an implicit sea-resort identity. However, by the 1920s there was a visible shift from a selling of health to a selling of recreation and the mid-twentieth century identity of Portrush was one of holiday-making, outings, swimming, dancehalls and amusement arcades (Byrne, 2005).

Figure 5.5 Ladies Bathing Place, Portrush, Co. Antrim, c1910
Source: Courtesy of the National Library of Ireland.

A Bath House was built close to the promenade and provided hot, cold, sea and seaweed baths in 1866 for the convenience of guests at the Antrim Arms Hotel but which were also available to the general public. Later in the century, separate bathing beaches were developed along the sheltered eastern part of the front, with the construction of a Ladies Bathing Place, complete with ramshackle changing booths. For the men, the Blue Pool was a popular swimming area, where naked bathing was commonplace until well into the twentieth century. There were also extensive beaches to the west and east. In the inhabitations and behaviours of bathing at Portrush, the role of the fresh air and clean water in place were extensively used with the 'pure, dry and bracing climate' used to attract visitors. Spatially, the production of place was shaped by local entrepreneurs and business interests in which local landowners and hoteliers played a key role. The development of a large 'signature' hotel, the Antrim Arms, was representative of similar practices elsewhere and the development of the external, and later internal, bathing facilities augmented the connection between water, health and place. The Northern Counties Railway Company was also an important agent in place creation, ultimately shaping access to the town. Indeed the social cachet of the town, when compared to its neighbouring resort, Portstewart, a few kilometres to the west, was shaped by the railways decision to complete the line initially into Portrush. This meant that visitors to Portstewart were obliged, until the completion of a branch line some years later, to get a horse and car to the latter town, something only the wealthier visitor could afford.

In built environment terms, Irish sea-bathing resorts contained generic beach side facilities such as bathing-boxes, slipways and functional (rather than commercial) piers and harbours. Additional entertainment facilities included promenades, parks and landscaped walks such as the Doneraile cliff walk at Tramore and the paths around Bray Head. Hotels were an essential part of the built landscape and in line with practices elsewhere were designed, in name and form, to provide metaphoric associations with place, health, luxury and grandeur. The semiotics of hotel names represented a sort of 'Nouvelle Marine' signature providing imaginative narratives of 'escapes to dreamland' (Lenček and Bosker's 1998). Bray, Kilkee and Dun Laoghaire all at some stage contained a Royal Marine Hotel, while the International Hotel at Bray was one of the largest in the world on its completion in 1862. In an echo of St. Ann's Hydro at Blarney, the same hotel had an ephemeral health identity as the Princess Patricia Hospital for wounded soldiers during World War One (Flynn, 2004). The Great Hotel at Tramore changed its name to the Grand while there was also a Hydro Hotel in Kilkee as well as hotels with affinitive place associations such as the Causeway at Portrush (Byrne, 2005; Hackett, 1994). The designs of these hotels did, for their time and especially locations, provide aspirational settings which incorporated at least some consideration of therapeutic design in their views, facilities and services, especially at the Slieve Donard in Newcastle and the International at Bray (Young, 2002; Davies 2007). As well as these more formal sites of accommodation, there was a plentiful supply of lodgings while at resorts like Youghal, Tramore and Kilkee, large and well-built villas provided the proverbial 'quality for the quality', meeting the demand for reliable accommodation for annual family visits. Heuston noted that the renting out of houses as lodges in Kilkee during the summer, while profitable, must have required the displacement of local residents, presumably to temporary stays with friends and family (Heuston, 1997).

In line with arguments emphasizing the importance of place in the performance of health, many of the Irish sea-resorts championed the therapeutic qualities of their physical locations. As many of the resorts were surrounded by attractive countryside, such narratives were developed by local and external business interests such as railway companies like the Belfast & Northern Counties Railway Company at Portrush. At Bray, discursive practices common at English resorts were used, whereby favourable statistics on climate and morbidity rates were used to sell the place (Clare, 1998). In addition, the town's location at the foot of Bray Head and proximity to the Wicklow Mountains were mirrored in the invocation of the therapeutic qualities of the Mourne Mountains at Newcastle, County Down (Knox, 1845; Davies, 2007). At Youghal, Knox noted the favourable aspects of place in its formation as a therapeutic setting:

> It is much frequented in summer on account of its beautiful level and extensive sea beach, mild climate, and agreeable and salubrious situation. It is sheltered on the west by a hill, and open to the southward in the direction of the bay, which is nearly land locked. (Knox, 1845, 251)

As affective place, the sea-resort was also sold on additional spatial qualities of safety and tranquillity. The contrast at Kilkee between the sheltered and protected beach and the wild ocean outside suggested a safe haven, while the role of place in rejuvenation was ascribed to 'the abundant supply of life-giving ozone which invigorates the old ... giving the invalid a new lease of life, and sends all back with a renewed energy' (O'Carroll, 1987, 26-27). Even in 1831 the stillness of the resort was specifically advertised by Mrs. Shannon's boarding house where; 'the most perfect order and quietness may be depended on' for invalid visitors. Therapeutic design was also evident in lightly coloured houses and large windows facing the sea to take advantage of views and the marine light, noted as a marked contrast to inland building at the time (O'Carroll, 1987). Within the enclosed confines of the Victorian seaweed-bath, at Enniscrone or Kilkee, where 'saline vegetable baths' were advertised in 1841, the embodied experience of a deep steaming bath while covered in oily green fronds was both restful and restorative.

The mobility of health practices and identities at the sea-bathing resorts were, as with other settings, expressed seasonally and temporally (Towner, 1996). The spiritual curative power of the August spring tides have already been noted while it was the tradition for no-one to enter the sea at Tramore prior to 1 May, the date of the pagan *Bealtaine* festival (Taylor, 1996). In the typical bathing resort, especially in the early periods of development, bathing and healthy activities were undertaken in the morning, social activities in the evening and a mixture of both in between. At Kilkee, de Bovet noted that the beaches were deserted at seven in the morning, noting: 'Men are forbidden to bathe after an hour so early that most prefer to go out some way along the coast, where they can enjoy themselves without infringing the regulations' (de Bovet, 1891, 186). Yet the men's exclusion from the safer beach during daylight hours, mostly for moral reasons as they traditionally bathed naked (and sometimes on horses), meant they had to swim in dangerous rocky coves like Burns's Cove and the Pollack holes which led to frequent drownings (Murphy, 1977; MacNeill, 1988). In Bray, towards the end of the century, there were a number of complaints about men wearing 'short bathing dresses after 9 am at the Cove while ladies passed by', though complaints about young ladies hanging around the male bathing-places were also recorded in Tramore in 1914 (Taylor, 1996; Davies, 2007). These regulated temporal and spatial performances were common across most of the larger towns studied, an arguably ironic identity given their enactments in places where time was supposed to be suspended. Shields and others note that this obsession with the regulation of time, may in part have been an displacement of work to leisure practices, but reflected the greater ordering of time in wider hydrotherapeutic settings as well (Shield, 1991; Mackaman, 1998).

The building of a range of different sea-baths reflected additional aspects of time and space regulation. While the building of Kelly's baths in Bray in 1861 helped a shift towards the sea, at other resorts, the preferred places for healing and leisure often meant a turning away from the natural location, the ocean, to the built setting of the lido or covered bath, for which there were nominally health-related explanations. Narratives of risk associated with the unpredictable and sometimes unclean sea as well as the presence at a number of baths of health-related treatments explained such spatial mobilities. At the Victoria Baths in Dun Laoghaire one could, at the turn of the twentieth century, enjoy a range of treatments and baths, while more vernacular connections in the twentieth century at Kilkee, reflective of the seasonal rhythms of the agricultural calendar meant that; 'Late in the season, after the wearying harvest months, farmers and their families came to Kilkee to benefit from Purtill's therapeutic seaweed baths which were widely acclaimed as cures for all kinds of aches and pains' (Heuston, 1997, 22). Yet as with other forms of hydrotherapies, the health identities of the sea-resort were mobile and ephemeral and while new identities of leisure and pleasure became normative, older narratives of health were never entirely lost. Even now new identities emerge at the beach in the on-going appreciation of the value of fresh-air, exercise and sports like surfing and sailing as continuing, but still wellness-related performances in place.

Economic Performances of Health: Designs and Appropriations

Sea-bathing resorts, in common with other therapeutic landscapes, were commodified in ways which represented distinct economies of health performance (Gesler, 1992; Geores, 1998; Hassan, 2003). As external demand drove local supply, there was money to be made at the seaside even before the wider social valorization of leisure time. The forms in which this occurred varied over time and health was variably involved by period and location (Corbin, 1994; Mackaman, 1998). As was the case at the spa town, there were entrepreneurial alliances at play in the Irish resorts, with landowners and capitalist entrepreneurs centrally involved in investment in place (Davies, 2007; Kelly, 2009). Using the tried and trusted metaphors of place-selling, health narratives were backed up by a clear expression of the social and material spatial conditions at the sites. Thus Knox recorded at Bray:

> The property of the Earl of Meath, is a more favourable bathing place than either Wicklow or Arklow, a preference to which it is justly entitled. It has an excellent hotel, every facility for hot and cold baths, a pleasant promenade to the sea beach, and a series of thatched cottages, kept with great neatness, which furnish accommodation to summer visitor, at a rent of £40 and upwards for the season. This cheerful little town is about ten miles from the metropolis, lying in a rich and improved country; but here, as at most other places of the kind in Ireland, we have to notice the want of suitable bathing machines. (Knox, 1845, 150)

Here the writer, himself an advocate of the water-cure, produced a promotional narrative which spoke of order, cure and facility, but which also invoked important colonial references to ownership, costs, accessibility and in line with other discourses, provided a dissonant health promotion whereby alternative sites were compared unfavourably to Bray (Towner, 1996; Kelly, 2009).

While the presence of the raw ingredients of sea-water, accommodations and amusements were all important in the successful development of the sea-resort, they also required an injection of entrepreneurial capital, an essential binding fluid in the mix (Walton, 1993; Hassan, 2003). At Bray for example, the period between the late 1850s and 1870 marked an especially busy period in the town's construction as a health resort (Clare, 1998; Davies, 2007). With investment by local landowners, the Earl of Meath and Viscount Brabazon and entrepreneurs/investors like Breslin, Quin, Putland and Dargan, new hotels were opened. Relatively high-risk health-specific investments such as the Turkish Baths, the Hydropathic Establishment and the Martello Terrace Baths were all built in this period (Clare, 1998). Yet with the exception of the sea-baths, most of the other investments failed or were resold within a decade and by 1870, the main investors had either left or narrowed their focus. Despite Geores' reasonable assumption that metaphor was at the core of the successful production of a watering place; it also required economic vision and a commercial underpinning (Geores, 1998). In more modest settings such as at Kilkee and Tramore, slower and steadier progress was made, while greater volumes of lower class visitors were needed to re-energize towns like Bray after 1870 (Davies, 2007).

These formative economic narratives were widely repeated in general guides such as Lewis's Topographic Directory from 1837 which recorded faithfully the reputations of places and narratives of utilization as a short-hand for fashionability.

Thus even small resorts like Lahinch, were described as; '... of late rapidly improved on account of its fine bathing strand ... and much resorted to during the season'. Conversely, and surprisingly presciently, it described the limitations and potential of Youghal as a sea-bathing resort:

> The town is much frequented during the summer for sea-bathing, for which it is well adapted, having a fine, smooth, and level strand extending nearly three miles along the western shore of the bay; but as a watering-place it is deficient in the accommodation of good lodgings, which might be easily supplied by the erection of marine villas and lodging-houses at the Cork entrance to the town, along the declivity of the hill, which would command a pleasing prospect of the bay, the strand, and Capell island. This would not only increase the number of visiters (sic) during the season, but induce many persons to take up their permanent abode in the town. (Lewis's Directory, (Vol. 2) 1837, 726)

Such villas and lodgings were subsequently built in exactly the locations suggested in Lewis's Directory, though not until 25 years later. In addition the quote suggests that to be considered a proper watering-place, a full ensemble of water, accommodation and entertainments were required. Lewis's Directory was in effect an early national tourist guide that echoed earlier (Cook's) and later (AA & Railway Company) uses of such narrative forms (Horgan, 2002; Urry, 2002). In addition at the end of the nineteenth century, a former employee of Cooks, F.W. Crossley became active in promoting Irish resorts in general under the auspices of the Irish Tourist Association (set up in 1895), the forerunner of contemporary tourism agencies. This expressly tourist role also traded on the watering-place emphasis on health and recreation. Indeed Crossley was involved with others across the British Isles in lobbying for an Act of Parliament ultimately passed as the *Health Resorts and Watering-Places (Ireland) Act 1909,* which allowed local authorities to raise local taxes to advertise and promote their places (Furlong, 2009). Authorities in England and Scotland were less successful, much to their chagrin (A. Durie, 2009, pers. comm., 3 October). Four of the study-towns namely, Portrush, Kilkee, Bray and Youghal availed of the opportunity. From a purely economic perspective, competition between resorts was relatively common. Concerns were expressed at Kilkee in the early twentieth century at a slight decline at the resort, which were perceived as being due to the relatively poor quality of the golf course, especially when compared to the neighbouring and competing world-class courses at Ballybunion and Lahinch (Heuston, 1997).

As noted by Durie, the promotion and significantly, the authorization of the product, in this case sea-water, was of a quite different character to that of the spa town:

> In authenticating the claim of any waters to medical virtue (a much-used word), no spa could progress to resort status unless 'proofed' by an analysis. By contrast, the seaside resort needed no such particular stamp of approval. What

mattered was the general endorsement of science and medicine given to salt-water treatments and sea bathing ... Of course, the provision of amenities helped popularization: baths, bathing machines, walks and wet weather amusements. (Durie, 2003a, 197-198)

Here geography was much more central as a commodifiable spatialized product than at the spa with the selling points focused on the therapeutic place in terms of its local climate, beach, amenities and accommodation. While some of these constitutive elements were as much social as salutogenic, together they formed a balanced narrative to maximize marketing reach. Although water was not as central to narratives of health commodification, it still assumed an important position and variations in salinity and sea air/ozone still mattered (Granville, 1971; Durie, 2003a). As early as 1784, the quality of the water was noted at Youghal: 'In the summer months great numbers come to Youghall, for the benefit of the salt-water, which is here to be met with in the greatest perfection', which, combined with the presence of early bathing-boxes, described as 'apartments on wheels', suggest a proto-infrastructure of health already in place (St. Leger, 1994). Narratives of health, though on the wane in Portrush by the 1920s, still described the air there as the most invigorating in the United Kingdom (Horgan, 2002). While the quality of the air has been mentioned previously, its ascribed position on top of a nominal 'health league table' was part of a deeper commodification of health in place.

A spatial appropriation of the healing reputations of other places was widely imitated in the selling of sea-side resorts in Ireland (Durie, 2003a; Andrews and Kearns, 2005). Brighton in particular was heavily invoked with its name, and by extension curative reputation, attached to the commodificatory identities of Bray, Tramore, Portrush and Kilkee. Portrush made some attempt at distinction by selling itself as the 'Brighton of the North', while Kilkee was the 'Brighton of the West' for a period in the 1920s (Bassett, 1886; Heuston, 2007). French connections were applied to Bray as the 'Irish Riviera', while Biarritz was invoked, again at Kilkee, in the 1920s and 1930s, quite possibly a politically motivated and post-colonial choice of imaginative connection in the early years of nation-building (Heuston, 1997). At Bray the associations were so strong that the Turkish Baths were compared to the equally Orientalist Brighton Pavilion, while the nineteenth century seafront by-laws were copied directly from Brighton as an even deeper discursive and regulatory connection (Davies, 2007). Like the example of the spa, such 'invocations of place' ran the risk of over-selling, so that the visitor drawn by the imaginative lure of Biarritz style casinos, sunshine and good food and transport might have found Kilkee somewhat lacking in its material realities (Hassan, 2003). But these invocations of place, common to many of the sites studied in the book, spoke specifically to the relationships between fashionability, class and the reproduction of specific metaphorical elements of those places. While the Irish visitor might never be able to get to Brighton or Biarritz, they were perversely sold on its imaginative and affective elements of cure, escape, liminality and adventure (Shields, 1991).

Fashionability was also a factor and was implicated in the cyclical success of various resorts. As elsewhere celebrity endorsement was extensively used in the selling of health in place. Dun Laoghaire was visited by Queen Victoria in 1849, or more specifically passed through by her en route to Dublin, yet the town lived out on the royal connections for years and named the local baths after her. Ironically her son, Edward VII visited and stayed in the town in 1904 but is rarely mentioned in place narratives. Bray was patronized by the Lord Lieutenant in its early decades of popularity while Kilkee made much of its visitors which included William Thackeray and a honeymooning Charlotte Brontë in 1854 (Heuston, 1997). In those towns which lay close to particularly scenic tourist attractions such as Bray, Portrush and Newcastle in County Down, many famous visitors would have passed through en route to the respective delights of the Wicklow Mountains, the Giants Causeway and The Mourne Mountains. At the latter site, the location and therapeutic joys were turned into popular song by Percy French while enjoying a stay at the Slieve Donard Hotel, which was also patronized by Charlie Chaplin and King Leopold of Belgium. While the invocation and endorsement of celebrities can seem trivial, it was, in different time periods, essential to the reputations of place. It also tapped into the associated psychologies alluded to above whereby one envisaged oneself in an exotic location surrounded by exotic people. These invocative connections are well understood by tourist geographers but are also relevant to health geographies (Urry, 2002; Gibson and Connell, 2005). At heart lies the crucial factor, more directly commodifiable than fashionability, of *consumer confidence*, upon which many contemporary social, cultural and economic narratives still depend. For the purposes of the health cure, whether in the sea at Lahinch or in the baths at Bray, their efficacy and patronage was also dependant on precisely such a psycho-cultural factor.

As at other therapeutic places, ownership and elements of structure and agency were important to the development of the sea-bathing resort. These included developers (the literal landowners) and accommodation entrepreneurs working at a variety of scales, from big hotels to small lodgings houses. These often worked hand-in-hand with entertainment entrepreneurs to develop a variety of additional facilities to keep tourists in place. At Dun Laoghaire, new sea-baths were built in 1828 by Mr. Hayes, the owner of the Royal Hotel, and this process was repeated less than 20 years later by John Crosthwaite, who built sea-baths north of the town beside the railway line at Salthill (Quinn, 2003). In common with many British resorts, local government became much more active agents in the latter half of the nineteenth century (Towner, 1996). Bray Town Council were actively involved in local schemes, especially in relation to the development of the esplanade, sea-baths and other town improvements linked specifically to health and safety such as sanitation and lighting. In this they assumed entrepreneurial and promotional roles as a public agency, after the early burst of private investment failed (Davies, 2007). The same occurred in Dun Laoghaire as the Urban District Council took over the Victoria Baths, originally developed as a private initiative by Crosthwaite, in 1898 and extensively re-developed them between 1905 and 1911. This assumption of an entrepreneurial and developmental role by the newly created local authorities was also evidence of a much more formalized production of place and a more explicit tourism-related narrative wherein seaside resorts were re-commodified using public, as opposed to private money (Hassan, 2003; Furlong, 2009). Other external agencies reflected colonial models and included a range of railway companies, such as the Waterford and Tramore Railway, completed in 1853 and the Belfast & Northern Counties Railway Company at Portrush. The involvement of investors like William Dargan, who as major shareholder in a variety of railway companies, was actively involved in the developments of Bray, Dun Laoghaire and Portrush, suggested that agency was also multiply networked in financial terms. Here entrepreneurship extended not just to the profits to be made at the healing resort, but in relational geographies of spatial networks connecting those places to their demand nodes (Curtis et al., 2009).

In considering the health histories of the sea-resort, the commodification of health, while important in the initial development of the towns, became less significant as a stronger narrative of leisure began to dominate after World War I. In considering how health was shaped in and by its economic performance, the ongoing power of the water was a persistent theme. The performances of dipping, bathing and swimming were augmented by the use of sea-water in the treatments of invalids, both in and out of the water. The efficacy of the cures were also sold at associated hot, cold and seaweed-baths, a practice which sustains to the present day at Strandhill, Ballybunion, Enniscrone, Newcastle and Kilkee. Indeed the booklet produced by Hugh Hogan at Kilkee to promote his baths included those classic elements of water-cure promotion including a chemico-medical listing of ingredients, treatments and cures as well as the well-established tactic of patient-biography mixed with celebrity endorsement. A Baron James de Basterot, for more than three years, a 'martyr to the tic doloureux ... experienced a vast deal of relief from acute pains in my limbs, by the use of your warm sea-weed bath' (Hogan, 1842, 29). In addition, Hogan's guide, in an echo of spa towns like Harrogate, produced a 'visitor's list' so that the place was sold on its habitus as well as its celebrity. Broadly similar tactics were used at the end of the century by Crossley though with a more explicit resort-wide tourism promotion in mind, rather than any individual business within it (Furlong, 2009). Yet as with any product, there was no business without demand, and the sea-resort, was sold as much on its liminal cultural attractions as its specific value as a site of health. Both however contributed to the marketing of a place which had a sustained and imaginatively proven reputation for well-being.

Summary: Baths and Benefits

Though there is some tendency now to underplay the therapeutic role of watering-places, and to see this as a cover for the really important business of pleasure seeking, the provision of health was a vitally important function of resorts. Georgian society was as obsessed about its physical and mental well-being as we are today, and for those feeling unwell, a visit to spa or seaside, if they could afford it, represented a major opportunity to obtain relief. (Borsay, 2000, 798)

While Borsay's quote is useful in flagging up a stubborn residual heath meaning at the sea-bathing resort over and above the social identities, there was a genuine mobility in how that therapeutic role was interpreted, understood and enacted over time. It shifted from natural to built settings as sea-water consumption and early plunges into the wild Atlantic waters at Kilkee were replaced by the tamer immersive settings of the tidally-filled, but crucially, safer sea-baths at Dun Laoghaire, Bray and Newcastle (Towner, 1996; Borsay, 2000). In time, curative promotion used less specifically water-based, though still water-formed, narratives of the benefits to health and well-being of sea-air and ozone (Corbin, 1994; Hassan, 2003). In addition, the health appeal of resorts was noted by Borsay as initially lying in: 'their capacity to combine traditional magico-religious elements of therapy with new empirical 'scientific' ones ... on the sheer availability of the cure, particularly when sea-bathing was 'discovered', and the relative freedom of control this gave patients over handling their personal health' (Borsay, 2000, 798-9) .

In the ability of the sea-resort to retain older spiritual and natural narratives of healing with the new 'magic bullets' of scientific medicine, the discursive placing of health at the seaside drew from a range of health paradigms. In Ireland these shifts of health narratives from traditional uses of sea-water to more freely available and personally negotiated engagements with its newly 'medicalized' benefits were visible at places like Kilkee, Lahinch and Bundoran. Finally, some of those more vernacular constructions, especially in the seaweed-baths of Enniscrone, Strandhill and Ballybunion, maintained an independent healing tradition alongside the more colonial models of health in place (Heuston, 1997). In this they marked parallel but less-reported health histories which flow through all of the sites in the book and connect vernacular performances of health to contemporary practices found in the modern spa.

While the later therapeutic identities at the sea-resort saw an emphasis on measurable health benefits, as in the publication of mortality rates at Bray, those narratives also began to depend more on place as source of health, rather than as might have been the case, the benefits of the water. In addition, the leisure expectations and commercial developments of the resort became more and more important and visits to Bray, Kilkee or Portrush were increasingly informed by implicit and culturally-produced experiences of thrills rather than treatment (Walton, 1983; Urry, 2002). These mobile discourses of health and leisure identities were also reflected in the shifting geographies of place as a range of new littoral forms of villas, piers, cheap lodgings, fairgrounds and campsites appeared alongside public baths, parks and promenades (Walton, 2000; Urry, 2002; Davies, 2007). While often stratified in social terms, either formally or informally, the public spaces of the sea-resort enabled a social mixing previously unknown in many countries which in part explained the associated liminal and marginal identities associated with these often uncontrollable locations (Shields, 1991; Lenček and Bosker, 1998).

Health performances were informed and shaped by such cultural, spatial and economic dimensions. In those narrative productions of place, social and cultural discourses on the benefits of sea-bathing, the consumption of sea-water and the benefits of sea-air, swimming and other forms of exercise, were generated from vernacular *dinnseanchas* traditions and imported via colonial productions of place (Logan, 1981; Duffy, 2007). Such narratives further emphasized discursive connections with embodied health whereby the unhealthy body, in its arthritic, rheumatic or poxed form, was treated in or through water. From the consumption of sea-water in Portrush in 1760 to the midnight swims of farmers and their wives in Ballybunion, those bodily connections were both material and metaphorical. Similarly, the discursive production of the British sea-bathing resort was replicated in the 'mixed-colonial' Irish setting, not just materially, but also in the social, moral and behavioural expectations enacted in place. At the Bray seafront or around the bays at Kilkee and Portrush, cultural performances of class were ritual and seasonal enactments of understood behaviours, often in marked contrast to local resistance and liminalities. They also reflected the metaphoric production of place through invocations of other more famous resorts in the same type of cultural co-dependence enacted by the Anglo-Irish at the spa town, a process which continued after independence. The therapeutic power of the littoral landscape through the waters and airs of the resorts, were reinforced by the creation of bathing narratives to sell place. Here the benefits of the resorts were commodified explicitly in metaphors connecting sea=health (Geores, 1998). Despite some small contestations and subtle admissions of drowning stories at Portrush and of risk (and risk-averse behaviours) at Kilkee and Dun Laoghaire, those assumed therapeutic benefits of place were widely reported. While the social narratives were dominant in different towns at different times, lingering health metaphors did not completely disappear.

Images of place are a constant negotiation between the expected and experienced, the real and imagined and can even be seen as pseudo-events, where the real and imagined merge into hybrid reproduction (Boorstin, 1964). Yet, it is through inhabitation and dwelling that such enactments can be best understood and essentialist assumptions challenged (Ingold, 2000; Urry, 2002). In their embodied form, those inhabitations were found in the contradictory freedom and regulation of the female bather at Bray, simultaneously liberated and enclosed by seafront micro-powers of regulation and surveillance (Shields, 1991; Davies, 1997). Curative and cultural performances in place were affected by setting, age, gender and social class as the poorer native inhabited different parts and seasons in Kilkee and Youghal, reflective of social and spatial segregations elsewhere (Hassan, 2003). But those lived performances were an essential part of place production, reflecting mobility and dwelling and the interflow of rest and movement, stillness and healthful activity to be found at the beach (Conradson, 2007). In this they also reflected Harvey's more affective understandings of landscapes as being formed so that:

> ... repeated encounters with places, and complex associations with them, serve to build up memory and affection for those places, thereby rendering the places themselves deepened by time and qualified by memory...To dwell, according to this kind of interpretation, is to become rooted by the act of accommodation in place. (Harvey, 1996, cited in Jones and Cloke, 2002, 81)

While these affective and internally owned aspects of 'dwelling' and place inhabitation are linked to material experience and metaphoric associations, they are ultimately formative in themselves in experiential terms of psycho-cultural formations of place linked to the noises, sights and affective memories of the seaside (Jones and Cloke, 2002) Just as Conradson (2005b) challenges assumptions of a solely positive identity of therapeutic landscape, so the individual experience of the sea-bathing resort can be both positive, as in the patient whose eczema was alleviated by repeated bathing at the Collins' Seaweed Baths at Ballybunion, and negative, as in drownings at Portrush and Kilkee (Collins and Kearns, 2007). It is in inhabited experiences that the therapeutic elements of the seaside watering-place are most fully enacted in embodied, performative and mobile form.

Finally, the notion of ownership is an important element in a deeper understanding of therapeutic performance at the sea-bathing resort. This psycho-cultural ownership is negotiable at a number of levels, from the individual to the structural. Just as Foucauldian notions of control and micro-powers of surveillance were enacted in Bray, Kilkee and Ballybunion, so wider elements of political economy were involved in cultural discourses of place production. In the solidly colonial setting, entrepreneurship and commodification of health at the seaside were, through individual investors and landowners like Dargan or Major MacDonnell at Kilkee, ultimately shaped by wider behavioural and economic models (Heuston, 1997; Davies, 2007). This was continued in a post-colonial Ireland by the particularly powerful link between the State and the Catholic Church, where moral geographies were also visible in the gender and social segregations on the seafronts as well as in place behaviours. At Kilkee, the local parish priest railed against 'flagrant immodesties in dress and in general conduct' in 1935 (Heuston, 1997). Yet within these constraints, the inhabitation of place reflected individual and communal ownership, in the continuation of local practices such as use of seaweed-baths and in wider liminal and resistive behaviours. In Bakhtin's image of the split upper (imaginative) and lower (real and often grotesque) bodies, psychological separations of mind and body can also be seen (Bakhtin, 1984; Shields, 1991; Ó'Cadhla, 1997). Here the controls of space and curative behaviour, in the dippers at Dun Laoghaire and the bathers at Bray can be contrasted in forbidden and allowed behaviours (Shields, 1991). Personal ownership of one's own embodied performance at the beach has always had a similarly mixed expression, as act of health and social production, which had an ongoing contemporary manifestation in the links to public health discourses of risk and safety. In broad health-focused narratives of the beach as sites where one can escape, rest, relax and restore, continuities into contemporary therapeutic landscape such as the modern spa are visible. Yet even here, as developed in the next chapter, older and deeper psycho-cultural performances of health in place provide a connective thread (Conradson, 2007).

Chapter 6

The Modern Spa: The Relaxation Cure

Introduction: Retreat and Restoration

> Upon arrival in the destination spa reception, one can see through to the gardens behind, where rolling folds of green sward and blue flowing water are incorporated against a backdrop of established trees. The buildings which curve around the south side are for accommodation, those on the north side are the spa space and they are connected via long, glassed corridors as if vessels within a living organism. Around this living space, silent white-bedecked creatures float by on fluffy slippers en route to the therapeutic waters of the spa suite. (Author's Notes, Monart Spa, 27 March 2008)

When uttered in modern conversation, the word spa seems to produce an almost instinctive affective response, especially in the female listener. A visible relaxation of the shoulders and a slightly pleasurable exhalation of breath mark an embodied response to an imagined place. In the destination spa at Monart that image is reflected in the blue water, the green landscaping and the white robes circulating in the space. These therapeutic associations between word and place are experienced in modern spas through a combination of relaxation and wellness within specifically designed spaces. Yet simultaneous feelings of pampering and luxury are also implicit in the term, which challenge those health and well-being identities. In this mix of constructed curative space and cultural practice a familiar set of themes associated with the 'watering-place' re-emerge. While the previous settings described in the book are primarily historical, they all have, even as relict features, some foot in the twenty-first century. It is useful, therefore to consider a contemporary therapeutic setting, the modern spa, to examine how older performances of health are reflected in contemporary practices.

The modern spa has become a cultural phenomenon world-wide since the early 1990's, mirrored in an increasingly affluent Irish society, with global identities as sites of relaxation, rest and luxury (Smith, 2003; Connell, 2006). They have associations with health and healing, especially in relation to CAM and stress that at times sit oddly with parallel associations with beauty and pleasure (Williams, 1998; Fáilte Ireland, 2007). These modern performances of wellness reflect the past, especially at spa towns and Turkish baths. As cultural spaces within Ireland, they are also informed by increasingly connected and globalized markets in such therapeutic forms (Smith and Kelly, 2006; Hoyez, 2007b). The reproduction of such place ensembles with health associations reflects Hoyez's work on yoga spaces as an example of the 'globalising therapeutic landscape'. They also raise important questions about authenticity, identity and therapeutic design but also point to earlier Orientalist performances of health (Gesler and Kearns, 2002; Hoyez, 2007a). The modern spa is a highly commodified version of a therapeutic landscape, visible in multiple forms. These range from health farms, rehabilitative communities and clinics, to commercially-driven hotels with limited spa functions and include the increasingly popular day-spa (Smyth, 2005; Smith and Kelly, 2006). The discursive associations of 'hotel and spa' have become particularly commonplace in modern cultural narratives (Smith and Puczko, 2009). As a wellness form, the 'spa treatment' also has a complex fixed and ephemeral geography ranging from the high street to airports and festivals (Lea, 2006).

Given the lack of biomedical 'certainty' about many of the healing benefits in all of the studied settings, modern spa entrepreneurs have attached their own meaning to, and often exploited the term, to sell health and well-being. An alternative reading finds this acceptable in allowing for a self/lay determination of the potential carative benefits inherent in the modern hotel and spa setting. While more medicalized in mainland Europe, there is limited engagement with modern spas from health agencies in Anglophone countries, suggesting that in wider public health terms, such settings are seen as having little therapeutic importance. This does not mean that such sites have no value as therapeutic landscapes, rather that their healing dimensions must be considered in more psycho-cultural terms (Clarke, Doel and Segrott, 2004; Smith and Puczko, 2009). In such a conceptualization, there is a genuine value in looking at how such sites act as places of restoration and recovery and reflect those earlier dimensions of the water-cure (Porter, 1990; Durie, 2003a).

As narratives of wellness and well-being are at the heart of the modern spa, health treatments and experiences are more expressly restorative and carative than medicalized and curative. From massages and body-wraps to Kneipp-walks, the forms of hydrotherapy enacted in Irish settings like Monart, Inchydoney and Delphi speak as much to the past as the future. While more recent CAM treatments such as reiki, rasul and stone therapies are also found, water remains significant within treatment regimes in such settings. Virtually all of the modes of transmission noted at historic watering-places, ranging from ingestion to introduction, re-appear in different forms at the modern spa. In addition, non-Western forms of treatment and medicine, especially Ayurvedic, make an appearance at a number of the spas, reflecting the globalized therapeutic landscape (Hoyez, 2007b). Idealized spaces, of mountains, water, temples and spirituality are represented by what Appadurai terms 'ethnoscapes', and partial expressions of such forms are found within Irish hotel settings as at the Yauvana Spa in Cork (Appadurai, 1996; Hoyez, 2007b). More significantly, such sites also exist as settings in which self-landscape interactions are performed, indoors and out, reflecting recent work around relational and experiential aspects of healing (Conradson, 2005b and 2007; Lea, 2008). Here individual therapeutic experiences are worked out in place, through a range of self-health-place connections, though some of these may be contestable in the perceived pampered settings of the modern spa.

Typically, the modern spa provides a range of treatments and services to the user while the growth of the spa and wellness business globally has been considerable and is currently worth billions of dollars world wide (Smith and Puczko, 2009). In their material form, modern spas are similar to older hydropathic establishments, through the recurrent use of sweat-cure forms like saunas and *hammams* along with swimming pools, baths, jacuzzis and other water-based therapeutic elements. Such elements also reflect the almost standardized production of place identities, increasingly franchized within global chains of consumption (Lee, 2004; Smith and Puczko, 2009). The settings vary though the more distinctive and specialist centres tend to be located in 'therapeutic' rural places. This is certainly the case in Ireland, where nine of the 10 spas studied were set in smaller resorts or rural locations. While most of the places discussed date from the 1990s or later, several are located in sites with older colonial and therapeutic histories, which speak to the vertical dimensions of health narratives and 'watering-place' pasts (Andrews and Kearns, 2005). While for some, the location of the spas in private hotel settings might seem opportunistic, they are continuations of earlier spas and sea-bathing resorts, where health-place identities were enacted in precisely such sites (Davies, 2007).

As a starting point, 49 different spas, listed on a central business website, were identified as possible sites (Spa Ireland, 2008). In identifying 10 sites, spa classifications, ranging from destination spas to a range of different hotel spa types, were used to represent different geographies and morphologies. The distributions of the original sites and selected place-studies are shown in Figure 6.1. The three main sites chosen were Kelly's Resort Hotel and SeaSpa, a long established seaside hotel in Co. Wexford (see Box 6.1); the Kingsley Hotel and Yauvana Spa in Cork City (see Box 6.2); and Monart, a rural destination spa also in Co. Wexford (see Box 6.3). Other spas included the Cliff House Hotel in Ardmore, Co. Waterford and the Slieve Donard Hotel in Newcastle, Co. Down. Both of these had older therapeutic identities augmented by new spa facilities. Other sites with colonial histories included Bellinter House and Bathhouse Spa in Co. Meath and the Farnham Estate Hotel and Health Spa in Co. Cavan. Three final spas chosen for inclusion were set in rural landscapes; Delphi Mountain Resort in Co. Mayo, Inchydoney Island Lodge and Spa (specializing in thalassotherapy) in Co. Cork and a second destination spa, Temple Country Retreat and Spa in Co. Westmeath. In most cases, interviews were conducted with the owners and/or managers of the spas between Spring 2008 and Autumn 2009. They were asked a range of semi-structured questions on ownership, histories, treatments, catchments and clienteles, as well as health specific question on therapeutic design, the place of health and healing and their interpretations of health in their own places.

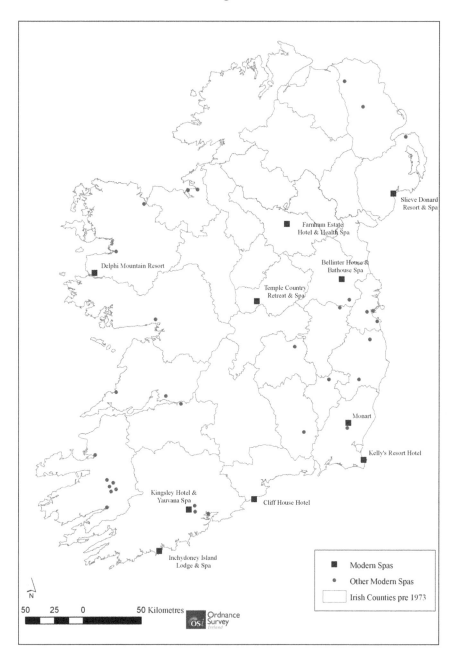

**Figure 6.1 Map of the Distributions of Modern Spa with 10 Place-Study
Locations**

Source: Spa-Ireland 2009. Boundary Maps. Ordnance Survey Ireland. © Ordnance Survey
Ireland/Government of Ireland Copyright Permit No. MP 008009.

Embodied Performances of Health: Robes and Reiki

In keeping with performances of health within other therapeutic landscape settings, the body remains central. As places in which bodies are enclosed, both in private therapeutic spaces and treatments, but also within the ubiquitous 'uniform' of towelling robe and slippers, a form of embodied 'wrapping' is enacted in a variety of ways. Similarly, the rituals of healing in place follow enactments elsewhere. In bodily movements through a variety of treatments, from hot to cold and back to hot, followed by time spent in carefully designed recovery rooms, older performances at the spa or Turkish bath are replicated at Monart or Farnham (Kelly, 2009; Shifrin, 2009). The experiences of health in place are also personal and relate to individual needs, conditions and expectations (Conradson, 2005b). Finally, these personal inhabitations of space are linked to how that individual relates to wider mind, body and spirit aspects of both the treatment spaces and their own affective responses (Tolia-Kelly, 2006). Whatever might be said, often in pejorative and disparaging terms, about the health metaphors and material forms used at the modern spa, it is in its inhabited dimensions that a fuller understanding of its psycho-cultural meaning as therapeutic landscape can be found. Holistic aspects of a CAM or even new age spirituality, which lie at the heart of some of the treatments, are expressed in a lived sense, even if there is cynicism around their specific benefits (Williams, 1998). It is for their imagined healing benefits, as much as verifiable cures, that performances of health are enacted, further reflecting the interplay of organic medicine/health and the psychosomatic placebo (Porter, 1990).

In their embodied performances, health-related treatments included a variety of massages therapies and water-specific forms such as baths, steam rooms and therapeutic pools. At Monart, the spa suite included *hammam* rooms, a walking Kneipp pool, a hydrotherapy pool, ice rooms and an outdoor sauna (Monart Spa, 2008). An assortment of balneotherapies were offered at Farnham, Delphi and Slieve Donard, while at Inchydoney, its treatments had a strong health discourse specifically focused on thalassotherapy (Inchydoney Island Lodge and Spa, 2009). Here thalassotherapy was invoked as a process of 'sea-water therapy', which 'allow the natural elements of the sea to soothe and relax'. Specific treatments at Inchydoney included warm hydrotherapy baths as well as back and neck massages. While most of the spas provided different massage treatments, often as part of a wider beauty package, their muscle relaxant and physiotherapeutic values were also emphasized. In addition several spas provided seaweed baths, reflecting older vernacular curative treatments for skin and rheumatic complaints. A fuller connection between the Voya Baths at Strandhill and the spa at Bellinter is discussed below, but seaweed baths were advertised as treatments at Delphi, Kelly's and Inchydoney. where algal wraps were also an option. These uses of historic and native forms of water-cure in the modern globalized setting of the modern spa also emphasized the sustenance of embodied performances of health in place.

Ironically, initial interview responses by the owners and managers at 'hotel-and-spa' sites, including The Cliff House and Kelly's, suggested that health was something that they did not sell, at least not explicitly. Such a view was however, shaped by the interviewees viewpoints, where health appeared to be seen in quite biomedical terms. This did not apply in all cases and one respondent stated;

> ... clearly health is about mind as well as body ... I suppose the site is also about rest and relaxing and has a fairly holistic edge to it ... there's a huge amount of evidence from health psychology about the need for this ... most of us have no time to reflect in this day and age so the site is laid out to help with that ... (Respondent M1)

This was a response from one of the destinations spas, where one would expect a stronger expression of wellness, especially when compared to the more hotel-based spas. A more consistent theme emerged around wellness narratives of rest, relaxation and retreat, linked to notions of de-stressing. At Monart, promotional material described it as; '*the* destination to rest, rejuvenate, transform' (Monart Spa, 2008). Implicit in these texts was the need to refresh but also to ready oneself to re-enter the world in an altered form. Retreat and travel to generally remote settings to re-find oneself echoed earlier 'health pilgrimages' to wells and baths (Urry, 2002). Such narratives were repeated at Bellinter, Farnham and Inchydoney, where at the latter, one was invited to 're-introduce your body to your mind'. This almost Jungian conflation of embodiment and place-based rejuvenation also suggested how the 'split' mind and body could be re-integrated in place (Philo and Parr, 2003). These holistic discourses were scattered throughout the promotional material of all of the sites, displaying clearly that narratives of healing lay at the heart of their metaphorical identities. As a final element, new public health discourses around risk and work-life balance, identify clear associations between health and stress with such psycho-cultural elements having biomedical as well as wider health consequences (Wilkinson, 2005). As a by-product of this, the need to de-stress is increasingly seen as a health-specific response and one which was fully embedded in the narratives of health at sites such as Temple Spa, that suggested; 'The spa practices have measurable physiological benefits ... a measurable change in hormonal balance that reverses the effects of stress on the internal chemistry of the body (Temple Spa, 2009)'. While full biomedical acceptance of such physiological proof is still at a developmental stage, the relationships between stress and a wide range of degenerative diseases such as cancers and cardiac conditions are increasingly well-established (Heller et al., 2005).

Box 6.1 Kelly's Resort Hotel and Sea Spa, Rosslare, Co. Wexford

Originally founded as a Tea Rooms in 1895, the hotel has been run by four generations of the Kelly Family since then (Kelly, Kelly and Foster, 1995). Unusually for a long-established hotel in Ireland, it has an identity which is relatively non-colonial, with a close place affinity to Rosslare and the beach. Connected to the rest of the country via a railway line built in 1882, the development of a golf course and proximity to cross-channel steamer traffic gave the hotel a prime location. As an establishment with a strong entrepreneurial place identity, especially for families and return visitors, the hotel has always had a national reputation. The most recent development has seen the development of a new spa, the SeaSpa, built around 2005 as a separate wing on the hotel, in place of the no-longer popular squash courts.

At the hotel, performances of health were until relatively recently, set within discourses of exercise and leisure, linked in part to the proximity to Rosslare Strand, but also based on the hotel's internal facilities. The development of the SeaSpa was noted by Respondent KL as being almost expected by the contemporary clientele. Yet it is also representative of a mobile notion of space reproduction, as the owner's development of the spa was motivated by visits to similar facilities and a recognition of a 'wider hotel and spa' metaphor at play across global settings. While health is not always fully central in place narratives, the development of the spa, using a mix of native and globalized treatments and products, is also reflected in other developments in the hotel, such as specialist health promotions, drawing in expertise from a group of holistic experts associated with RTE, the national broadcasting organization. The development of the SeaSpa also marks a spatial segregation from the older and noisier swimming pool facility at the other end of the site, where a separation of public/active space from private/still space is apparent. In addition, the spatial invocation of the wider therapeutic setting is enacted through the pumping of sea-water into the spa, as well as the use of maritime imagery in the outside spaces which connect and link the spa building to the nearby beach and sea. The involvement of an international spa design company is also representative of a wider globalization and reproduction of modern spa space, though at Kelly's the present owner was insistent on some variations being introduced to reflect the location of the spa on an Irish beach, through the use of specific materials and designs. The long-standing reputation of Kelly's as a family hotel sees the spa as a relatively recent innovation, and interviews noted the opportunity for traditional users to explore a set of new wellness performances. While there are some tensions in this co-location and the shifts in new codifications of health/place, the affective connections to Kelly's as historic site of rest and restoration continue to be enacted in recurrent and transformative inhabitations of the site.

Figure 6.2 Grounds and SeaSpa, Kelly's Resort Hotel, Rosslare, Co. Wexford, 2008

In talking about alternative health practices and wellness treatments at the modern spa, there are a number of accepted CAM forms, especially ayurvedic and traditional Chinese medicine (TCM) that are extensively practiced worldwide (Heller, 2005; Hoyez, 2007a). Within the chosen Irish settings, the Yauvana Spa, at the Kingston Hotel in Cork, is of particular interest. In this newly-built hotel spa, a separate ayurvedic treatment centre has been set up to provide specific health services and treatments with its own trained doctor and five ayurveda-qualified practitioners from India. On one level this acts as a post-modern reversal of recent trends in health tourism yet at a health-place level, the spa strongly reflects the globalizing therapeutic landscape (Connell, 2006; Hoyez, 2007b). The presence of such treatments at an Irish spa is less surprising when considering that ayurvedic treatments are found not only in old spa towns such as Bad Homburg, but also extensively across modern spa settings in Europe and North America. Indeed ayurvedic treatments are also provided at Farnham, Slieve Donard and Temple, though not to the same degree as the Yauvana Spa. Temple also provides a further link to globalized medicine through the use of acupuncture. Even accounting for the partially Orientalist and trendy nature of such treatments within a globalized market of health consumption, these examples point to the variety of their spatial expression (Curtis, 2004; Smith and Puczko, 2009).

Embodiment is at the heart of the different performances of health, and a central metaphor for healing and restoration at the modern spa. Embodiment is not just a physical but also a mental and imaginative process (Smith, 2003; Timothy and Olsen, 2006b). The understanding that embodiment has a link to the mind and the spirit recognizes an emotional and affective dimension to the modern spa as healing place (Tolia-Kelly, 2006). While in some cases emotional responses to place are reflected in a retreat into one's own personal space, parallels might be drawn to the reflective and retreat-like spaces of the modern spa, albeit in a different and contradictory public setting (Davidson and Parr, 2007). Aspects of Conradson's notional 'geographies of stillness' can be recognized, where one's affective and personal responses to settings almost church-like in their peacefulness, silence and behaviours, are experientially enacted (Conradson, 2007). The hushed and consciously spiritual settings of the modern spa, bear witness to Knott and Frank's suggestion that they act as, 'secular sacred places', and provide a contemporary analogy for the shifting spatial metaphors of religion and health (Knott and Franks, 2007). Indeed Respondent SI suggested that massage is the physical equivalent of confession. Further embodied performances are evident at settings like Monart, where visitors were encouraged to cast off their clothes on arrival and spend the bulk of their stay in bath robes and slippers, a practice common within the confines of the separate spa facilities at Bellinter, Slieve Donard and Farnham.

In addition, the value of treatments such as body wraps, speaks to an opening up of the body, both physically and emotionally. In a valuable affective comment, Respondent SI noted the particular psychological value of touch that reflected wider emotional geographies research on touch as an aspect of emotional healing (Sibley, 2003; Paterson, 2007);

> There is a strong sense of indulgence and the sensual aspect of the treatment is quite important. In fact for a lot of people who go it may be the only human touch they get and humans need that you know so that's a selling point. Having come back from the US, we in Ireland are fairly out of touch with our bodies so this is important. (Respondent SI)

Embodied aspects of the cures, through bodily immersions in water, steam or mud also reflect older therapeutic practices. At Monart, the use of the caldarium, sanarium and steam room, as part of a notional *hammam* experience, mirrors the designs and bodily experiences of the Turkish bath or in a less luxurious setting, the sweat-house. In addition, taking a seaweed bath at Bellinter, wherein one first enters a heated wooden box sauna, prior to immersion in a bath containing natural seaweed, which clings to and oozes around the body, provides intense and embodied experiences which can be simultaneously off-putting and relaxing.

While it would be easy to suggest that there is little of 'health' value in a mud-bath and a facial, it does raise internal and external contestations of the place of health and wellness in the modern spa setting (Clarke, Doel and Segrott, 2004). Within the sites, the explicit position of a healing/wellness narrative is more closely associated with destination spas, whose identities and indeed visions reflect that wider holistic vision of health. At Temple and Monart, this vision is more apparent than in the hotel-based settings, where the specific health narrative of the spa is often ancillary, rather than central, to its metaphoric identity. At Bellinter, Respondents B1 noted;

> ... well we're not really a destination spa ... I think we are classified as a
> 'bijou spa' so the focus is on the more general spa break experience ... so as
> in the previous answer I think that health is not at the centre of what we do ...
> (Respondent B1)

This response is representative of confusion in the discursive place of health and
even a slight lack of confidence around its normative definition. This may be in
part linked to the contestations around health and beauty, but may also reflect a
contested understanding of health in the more-business oriented functions of the
hotel spa setting. Here the creation of a wellness-focused place identity is much less
confidently developed, and perceived to be less significant, than at the destination
spa. Despite the limited focus on health, almost all of the places studied have a
healing identity, often more explicitly used than the places themselves recognized.
At Kelly's Resort Hotel, the SeaSpa brochure specifically notes;

> Step inside and feel at once the serenity, calm and warmth of welcome; you have
> arrived at 'SeaSpa' in Kelly's Resort Hotel – a new concept in total well-being.
> Here, healing seawaters, heat and steam experiences blended with therapeutic
> lighting and textured surrounds will help service the body, mind and soul.
> (Kelly's Resort Hotel, 2009)

In addition, the hotel's visits from the RTE Health Squad (see Box 6.1) provide
advice on nutrition, holistic practices and fitness. This health promotion and
awareness role is again under-played by the hotel, at least in its own envisioning
of having any sort of health identity. This lack of confidence and belief in a wider
health narrative arguably reflects a wider lack of confidence in non-biomedical
health identities in society more generally (Doel and Segrott, 2003).

A more holistic health also lies at the heart of self-landscape interactions
found at the modern spa (Conradson, 2005b; Lea, 2008). As these interactions
are embodied, felt and experienced in place, phenomenological connections
between an individual or group of individuals are further deepened by its medium,
water. Here water as a connector between the natural landscape and the human
respondent is reflected in a whole range of healing terms which are scattered
across the brochures and promotional material of the modern spa. Words such
as; rest, restoration, relaxation, rejuvenation, revitalization, reflection, restoration,
rebalancing, refreshing (and that's just the Re's) are regularly utilized in 'naming-
as-norming' narratives that metaphorically connect healing and wellness to the
spaces of the modern spa (Berg and Kearns, 1996). In adopting a more restorative
and holistic position on what embodied health means in place, the necessity of
justification and authentification, in rational terms, is replaced by a more affective
and relational concern with rest and the restoration of balance in the body and
the mind. Indeed when seen in a wide range of bodily performances, it is in the
expression of inner tensions and mental strains that a form of healing takes place.

In allowing the body to be touched and caressed either by fluid natural or human contact, a curative engagement takes place.

A second contestation, around the more cosmetic and beauty related performances at the modern spa is also evident. While the beauty dimensions of well-being present a particular challenge to and draw a strongly negative response from, medical and social perspectives, it is something which is part of a wider consideration of mind, body and spirit. In particular, when seen in the light of embodied aspects of cosmetic restoration, reconstruction and repair, the mental health benefits cannot be discounted. However it should also be noted that apart from treatments for unsightly skin conditions, such interventions are associated with more medicalized spaces and settings (Connell, 2006).

Cultural Performances of Health: Privacy and Pampering

Health and wellness in the modern spa are also critically informed by social and cultural performances (Smith and Kelly, 2006). Irish spas clearly reflect globalized cultural productions informed by social tastes, cultural beliefs and material expressions (Heller, 2005; Smith and Kelly, 2006). Outside of European countries such as Germany, Hungary, Poland and Italy, where spa medicine cultures remained strong through the twentieth century, the original spa phenomenon had faded by the early decades of that century (van Tubergen and van der Linden, 2002; Urry, 2002; Coccheri et al. 2008). Yet stemming from social trends, such as the counter-cultural revolution of the 1960s, an interest in alternative and natural lifestyles was replicated in health terms (Clarke, Doel and Segrott, 2004). The increasingly globalized nature of that interest was augmented by travel, increased income and migration patterns, whereby different forms of health were either discovered, rediscovered or relocated by increasingly diverse and curious consumers (Wiles and Rosenberg, 2002). That such a trend coincided with a contestation of orthodox scientific medicine, via concerns around iatrogenesis, morality and ethics, deepened cultural interests in CAM (Porter, 1999). As both Illich and McKeown linked some of their arguments against hegemonic scientific medicine to water, it was inevitable that interest would be renewed in its healing powers (McKeown, 1979; Illich, 1986).

As a result, the spa, as both treatment and business, emerged from the 1980s in the US and UK, while in those countries like Russia or Japan, where spa cultures were still strong, modernized forms associated with tourism settings such as hotels and specialist centres were reintroduced (Petroune and Yachina, 2009; Rátz, 2009). What is important to note here, is that the growth of the 'modern spa' was driven by strong social and commercial interests where wellness and profit were co-dependent. This was also marked by shifts from the more social service models still extant in Germany, Italy and Hungary to privatized and commodified forms, often sharing the same settings and historical locations (Smith and Kelly, 2006). In Ireland, the increasingly wealthy and multi-cultural 'Celtic Tiger' consumer had come to expect the same provision of services at home as they might experience abroad. It can be argued that what emerged was a form of cultural neo- or globo-colonialism, expressed in the discursive production of new spa identities wherein health and social meanings were equally prominent. At the Yauvana spa this was especially the case, where Respondent KG1 noted:

> when we were starting the hotel there wasn't at the time much interest but ... the owners had been interested in it (ayurveda) ... and had travelled and (experienced it) and brought back the idea with him ... from a mix of business and then personal trips (to India) to find out more.

A similar link between travel and the bringing of ideas for spa formats and designs occurred at Kelly's in Rosslare where the hotel's owner had attended a conference at Cornell University and brought back the idea that a modern spa had potential and would also meet an incipient consumer demand.

As with the other hydrotherapeutic sites, the social activities to be enjoyed in and around one's water treatment remain an important part of the process. While one could assume that the hotel spa would have a socialized dimension, this applies to the destination spa as well. At both Monart and the other destination spa, Temple, the availability of high quality restaurants is promoted while both provide gyms for more active aspects of health and fitness. Temple also advertises the availability of good horse-riding and fishing in its vicinity, though not run specifically by the spa. Elsewhere, the range of ancillary activities depends in part on geography, especially where spa's such as Kelly's, Slieve Donard, Cliff House and Inchydoney are close to the sea. At each of these, traditional hotel facilities such as swimming pools and bars are augmented by local activities. At Inchydoney, proximity to a good surf break mean that it is advertised as an activity, while a range of sporting activities such as sailing and wind-surfing are also linked to Slieve Donard. In the case of both Kelly's and Slieve Donard, and reflective of earlier sea-resort identities, there are strong historical 'links' links to high quality golf courses. In all of the ancillary activities, the notion of sugaring the pill is revisited, though in stark contrast to the 'medicalized punishments' of the Victorian hydro, at the modern spa it is almost all sugar and very little pill. This balancing out of health and social functions affects their identities and viabilities as healing places. Too much pill embeds a health identity but makes them less attractive to consumer demand; too much sugar on the other hand makes them popular and successful yet is seen as diminishing their curative/carative value; a tension in many CAM products and treatments (Williams, 1998; Smith and Kelly, 2006).

Box 6.2 Kingsley Hotel and Yauvana Spa, Cork City

Originally the site of the Lee Baths, the public baths for residents of the west of Cork city, the site was redeveloped and reopened as a five star hotel in 1998 by two local hoteliers. The hotel itself was built from scratch on the banks of the River Lee and its initial health facilities were linked to leisure and exercise in the form of a gym and swimming pool for guests and outside users. However the big innovation in the hotel was the opening of the Yauvana Spa in 2006. While not part of the original vision of the hotel, the owner's had travelled themselves to India on business and holiday and were interested in the possibilities of introducing an ayurvedic spa to Ireland. Narratives of authenticity were important in the commodification of the spa, enhanced in the case of the Kingsley by the recruitment of a spa doctor and other staff directly from India (Urry, 2002). While some surprise might be initially expressed on its value and meaning in an Irish setting, Respondent KG1 suggested that there was a young (people in their 30s and 40s), enthusiastic and most intriguingly, informed clientele, who had had experience of ayurvedic medicine while travelling or living abroad.

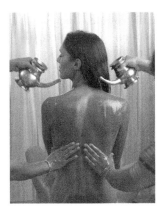

Figure 6.3 Extract from Yauvana Spa Brochure, Kingsley Hotel, Cork City, 2009
Source: Reproduced by kind permission of The Kingsley Hotel, Cork.

Within the Yauvana Spa, discourses of health and indeed medicine are explicit and central to its identity as therapeutic place. All patients are obliged to have a consultation with the spa doctor, who is fully-qualified in Ayurvedic medicine. She conducts this consultation to establish the specific mind-body constitution or *dosha*, which determine the patient's subsequent suitability for different forms of treatment. Those treatments are a range of massages, herbal remedies, stretches and oils which are strongly embodied, tactile and employ additional techniques of wrapping and beauty. Affect is an important part of such treatments, carried out as they are in darkened rooms, with strong sensual connections to wood, candle-light, incense, aromatic oils and human touch. Water is a significant element within Ayurveda, but not so strongly associated with the specific treatments, other than as a base for the oils and muds used in the massages. The accompanying and associated thermal suite reuses internalized narratives of sweat and detoxification familiar from many of the other watering places;

'Adjoining the spa we have a luxurious thermal suite with nine different exotic heat systems, designed to complement the healing properties of Ayurveda. The value of sweating for cleansing, detoxifying and relaxation is a documented accompaniment to Ayurvedic well-being in ancient Sanskrit texts dating back to 500 B.C. Today, you can enjoy them in the salubrious surroundings of Yauvana Spa' (Kingsley Hotel, 2009).

Wider notions of therapeutic design were incorporated into the spa and thermal suite as well as in the hotel more widely. While the final shape and appearance of the treatments were left to the Indian staff, the original interior space was designed by a UK based spa design company. Externally, the setting by the river was attractive to the owners, though the site has aspects of health risk as well in its location within a flood-plain.

Within the spa spaces, the roles of image and metaphor are central to culturally-informed performances of health. In virtually all of the brochures, the use of conventional images of infinity pools, relaxed clients in towels or robes, massage settings with candles and mood lighting all contribute to a form of place 'incensification'. At Monart, the brochure displays an attractive couple, mostly clad in toweling robes, in a variety of settings such as the pool, garden or in treatment rooms and relaxation spaces. These literal imaginative productions of place are expressly designed to act as metaphors for wellness and relaxation, a process usually augmented in texts encouraging clients to, 'be yourself again' and 'escape the stresses of modern life' (Monart Spa, 2009). These images then become part of the discursive practices and performances carried out once people get to the spa. Wandering around a public setting in a dressing gown is, in these settings, allowed, encouraged and even expected. These combinations of hint and regulation create a very specific performativity in the quasi-spiritual spaces, reminiscent of wider health geography concerns with stillness and retreat (Conradson, 2005a and 2007). In the designed spaces of Monart, where separate therapeutic rooms were connected around a central space, users negotiate their own movements, guided by notices and staff in the spa area.

Yet what cannot be ignored is that for many users, the social is at the heart of the spa identity. There are strong associations with pampering, relaxation and luxury and modern cultural understandings of the meaning of the word spa shape their image and identities (Lee, 2004; O'Mahoney, 2006). While healing and wellness narratives feature in the brochures and web-sites, they are just as readily interpretable in their appeal to the 'worried well', who are sold on the new cultural metaphor of Spa=Pampering and Luxury (Geores, 1998). Within the brochures produced at such settings within Ireland, those broader metaphors are fully visible, though it must be said that at Monart, Kelly's and The Kingsley, these are well balanced with narratives of a mind-body-spirit form of well-being. On one level, it is the glossy brochures which lead to a contestation of health efficacy and benefit, in the eyes of both the public and health professionals (Connell, 2006). Yet the glossy brochures are explicit elements of a cultural construction of place identity where the right buzz-words, such as pampering, indulgence and the spoiling of oneself, are carefully placed to reflect that identity and sustain the business.

These discourses are common to modern spa settings all over the world, and such luxurious aspects are significant to that identity, especially heading into a global financial downturn, where these kinds of expenditures are increasingly coming under scrutiny. The first health farm in Ireland, at Powerscourt, closed in late 2008. This may indicate that settings with specific health identities are as much at risk, if not more so, than those with combined curative and social identities (Cronin, 2008). Indeed the timing of the interviews has also reflected changing discourses of health at the spa. In the summer of 2009, Respondent SD suggested that clients were much more likely to see, or more specifically, justify, spa visits using the word health. This would have been unlikely a year previously as conspicuous consumption was expressed in a 'pampered and proud' narrative. Clients at Inchydoney in 2009 expressed a preference for more medicalized identities for treatments rather than plainer terms, a comment which suggests both a self-justificatory but also biomedically aspirational meaning to the spa experience. In these culturally-informed performances of health, one begins to enter into notional moral geographies of choice, guilt and consumption; a work in itself.

In such negotiations, the relationships between spa metaphors and individual spa experiences can be contradictory, being exploratory and directed, free and conditional, pampered and necessary. In contrast to the more self-directed and generic enactments at other spas, the performances of the Yauvana Spa at the Kingston Hotel are more specifically supervised and medicalized. Upon arrival a consultation is suggested with the Spa Doctor, who, under the principles of Ayurveda works with the patient to establish their 'physical and mental characteristics and ... pulses to determine their natural mind/body constitution or *dosha*'. This embodied patient-specific process is then used to determine which treatments are required, which were then followed by recommended usages of the thermal suite, again using specific metaphors of water-based health. In the embodied and directed experiences of Ayurveda one can see a specifically metaphoric and material use of touch and massage, performed in an 'exotic' setting by 'exotic' practitioners, with discursive links to globalized and other health identities (Vickers and Heller, 2005). Even at other less medicalized spas, such as Farnham and Slieve Donard, similar treatments are available and experienced in consciously affective settings where low lighting and the intensive use of candles and aromas provided a further example of 'incensification'.

While the more commercial cosmetic aspects have been noted, liminality does not appear to be a feature of the modern spa, when compared to older watering places. In their 'secular sacred' dimension and the fact that they are generally in private regulated spaces, the social behaviours are arguably the tamest of all the sites, including the holy well. One can see this as a perverse form of moral geography, wherein the visitor to Monart or Temple, especially those who are explicitly motivated by wellness and a holistic vision of health represent an almost paradoxical position. On the one hand, as consumers of CAM and alternative treatments, there are discursive associations with exploratory and open psyches, and a positioning as 'other', outside the mainstream of conventional medical beliefs and ideas. On the other hand, the experience and performance of these beliefs are set within strong moral strictures where balance and control, of the body, diet, exercise, consumption and behaviour can be as rigid and judgemental as a mainstream religion. The wider associations between CAM use and higher levels of education and income can also be seen as a potential, though arguably inadvertent, form of exclusion (Wiles and Rosenberg, 2001). In addition, earlier liminal connections between treatments like baths and massages have been to an extent de-sexualized and sanitized in the modern spa. Indeed the notionally repressed sexual Irish psyche was mentioned at Inchydoney (and at the seaweed-baths at Enniscrone) with Respondent IN noting that having a massage was no longer considered 'slightly sexual'.

In tracing the extent to which the health identities of the modern spa are affected and shaped by cultural performance, they strongly reflect the wider contested histories of the watering place (Urry, 2002). Cultural understandings of the spa in 2009 have much in common with 1709, reflected in metaphorical and inhabited elements such as expectations, experiences and performances. In the baths and pools, there are echoes of older forms of treatment, in which performances of immersion and expulsion remain the same. Yet these cultural performances of wellness also have, in their settings and human agents, a strongly socialized and culturally-formed expression. Even sites such as Kelly's, a traditional family hotel successful on its own terms, felt the need to embrace the 'spa revolution'. Respondent KL noted that part of the motivation to develop was simply that 'people expect this kind of thing now', so that the discursive creation of a 'hotel and spa' becomes a self-fulfilling act in an almost hyperreal sense (Urry, 2002). Here notions of habitus, in the sense of a collective class expression of taste are visible in an imitative cultural performance (Bourdieu, 1977). In addition while most commentaries assign a strong upper/middle-class identity to that expression, it can be, as shall be explored further below, more complex and less essentialist than that.

Spatial Performances of Health: Glass and Geomancy

Two broad areas of health geography research; therapeutic design and accessibility and utilization, are associated with spatial performances of health at the modern spa. In their locations in rural and 'natural' settings, as well as in their built form, notions of therapeutic design are materially and metaphorically embedded (Gesler, 1992; Curtis, 2004). All of the modern spas have a combination of design features, ranging from the materials used (such as glass, wood and of course water) to the architectural constructions which focus on therapeutic aspects, settings and combinations. Monart, for example, is designed around geomancy principles of curves and shape using local wood, while the spa rooms at Slieve Donard face out across the sea with a view of mountains from the interior spaces. As a second spatial element, the catchments, clienteles and locations of the modern spa speak to the same themes of accessibility and utilization as noted at other settings. In terms of wider histories of place, assumptions around clienteles and motivations for utilization can also be critically examined. Here fluidities of meaning and intent can be examined which connect the modern spa to other forms of watering place. The image of the modern wellness tourist, often characterized as affluent, female and interested in the sugared pill of treatment/luxury, can be both confirmed and contested in the modern settings of Temple, Kelly's and beyond. In addition the more private identities of the modern spa contrast with some of the more public facilities found elsewhere, and speak to common but complex threads of exclusion and exclusiveness which reverberate in watering-places across time (Kearns and Gesler, 1998; Williams, 2007).

Box 6.3 Monart Destination Spa, Co. Wexford

As one of only two designated 'destination spas' among the places studied, Monart's material form and development has a strong connection to holistic health identities. Based on a strong vision of place, the search for the site included the boring of a series of wells to provide locally derived drinking water, high in detoxifying potassium, within the spa. The site, in a townland called The Still, was originally an Anglo-Irish colonial mansion and while the main block of the original eighteenth century was maintained, the rest of the site was developed between 2003–2005. The design places an accommodation block along a curved western arm, while the thermal suite and restaurant facilities are on the eastern arm. There is a central reception area connected to the original house which contains meeting and library spaces. Therapeutic design is integral to the material production of the spa, with geomantic concepts of curves and orientation being expressed in the shape of the buildings. Externally, the wings face into a central garden space which containing water features including lakes, streams and small waterfalls. Created by a local designer in conjunction with the owner, the garden employs metaphors of stillness and movement and uses Celtic patterns while also employing native trees and shrubs, in a deliberate reversal of the exotic species introduced by the former colonial masters. In a further nod to local narratives of spiritual healing, the well was blessed and mounted by a statue of the Virgin Mary from the owner's mother's home.

Figure 6.4 Exterior Gardens, Monart Destination Spa, Co. Wexford, 2008
Source: Reproduced by kind permission of Monart Spa.

Figure 6.5 Internal Relaxation Space, Monart Destination Spa, Co. Wexford, 2008
Source: Reproduced by kind permission of Monart Spa.

Internally, metaphors of healing and wellness are most fully visible in the thermal area which contains a number of specifically hydrotherapeutic treatment spaces. Reminiscent of Turkish baths and hydros, facilities include a hydrotherapy pool, caldarium, steam-room as well as external and internal saunas. Evocative names such as *hammam* and Kneipp are also used while dark and light relaxation rooms provide contrasting affective spaces of recuperation. Narratives of wellness, in which water plays a major part, are reflected in the brochures and images of place with references to rest, restoration and rejuvenation and a 'safe haven' from the stresses and strains of the outside world. While such messages are both aspirational and experiential, wider inhabitation is expressed through a range of embodied and partially regulated performances. All patrons are encouraged to cast off external burdens through embodied acts of undressing and the wearing of dressing gowns and slippers. These symbolic acts of shedding and wrapping suggest, as do pods in the relaxation room, processes of re-birthing and the assumption of new or revived personas. Within the spa, conversation, outside of the communal eating areas or private rooms, is relatively hushed and this almost spiritual performance can be seen as a metaphor for a form of new age healing spirituality, even if such intent cannot be assumed for all clients. While never likely to be the most accessible spa in financial terms, the site strongly reflects the holistic intent of its owner and speaks to a modernist psycho-cultural ownership of health set within the individual's experience of place.

Specific therapeutic landscape elements of the modern spa are embedded in their designs and layouts, both external and internal. In particular the notion of spatial appropriation, as invocation and encapsulation of the settings and surrounds of the place, is fully played out at the modern spa. At the Cliff House and in Kelly's, proximity to, and views of, the sea are used to invoke water in a wider therapeutic sense as part of the discursive ownership of a healing space. In Kelly's this is visible in the use of maritime imagery in the design of the gardens around the

SeaSpa, where paths and the dunes on the beach edge have been used, along with the careful placement of buoys and nautical ropes as a 'Cape-codification' of place as land/sea interface. Indeed the nautical theme led to Respondent KL noting visitor comments that the gardens and new spa looked like 'a cruise liner just docked here on land'. The role of the sea as adjunct therapeutic place was fully expressed in the Cliff House brochure:

> Picture a perfect place. An intimate, cascading hotel where terraces, gardens and balconies seem literally to trickle down the cliffs. Picture a holiday and a hideaway. A hotel as distinguished for its design and dazzling sea views as for the region's finest food, spa and sleep experiences. (Cliff House Hotel, 2008)

Respondent CH, noted that they explicitly see the broad sweep of Ardmore Bay as an opportunity to 'bring nature into the property' while also noting that 'we do see the bay and the outside as our estate'. These connections between place appropriation and its discursive use as spatial healer also speak to new dimensions of a therapeutic landscape-as-gaze, when linked to affective spatial ownerships (Wylie, 2007).

The interiors of the modern spa also provide material and metaphorical evidence of therapeutic design and planning (Curtis, 2004). Water, along with stone and wood, are important elements in this interior space and at Monart, the Kneipp pool is augmented by a specially-designed hydrotherapy pool. Therapeutic design is also built into internal rooms where rest and relaxation are 'placed' into the post-bath recovery room at Bellinter, facing westward across a rural landscape. There are both light and dark relaxation rooms at Monart. Here individual or communal repose is enacted in affective settings whereby immersion in a dark space cuts out all external stimulus and forces the user inside themselves. In Monart's other communal space, large windows allow clients to recline and gaze on the carefully designed external gardens while set into the wall behind are 'pod seats', which in their discursive design form almost womb-like spaces of rebirth (see Figure 6.5). At Monart, affective, localized and even spiritual aspects of therapeutic design are deliberately incorporated;

> I also designed the site in way whereby there are no straight lines – a type of geomancy – so that the buildings and gardens are curved and circular. The garden is also designed on those sorts of principles. There are no mirrors or glass opposite doors either. … In the garden I also made sure we have some natural bog oak, as an example of local materials and the ironwork is designed around a blackthorn motif … there's also something about bringing the (natural) outside in to the building (Respondent M1).

At the Yauvana Spa, specific ayurvedic design principles were associated with colour and the use of specific woods and antiseptic oils. The same design elements to be found in the modern spa have echoes in medical settings, especially those associated with recuperation and recovery. While the latter may be more clinical than luxurious, such blending and merging of interior design, as material and therapeutic elements, are reflected in these privatized spaces of health and wellness (Kearns and Barnett, 1999; Williams, 2007).

The spatial locations of the spas are also relevant to any discussion of accessibility and catchments. A number of the spas are in remote (Delphi) or semi-remote (Monart, Inchydoney, Temple) parts of Ireland. Given that the choice of transport is invariably the car, these locations are less of an impediment than in the past. However, relative proximity to a large city is built in to the choices of location for other spas. Closeness to Dublin is important for both Bellinter and Monart, while Slieve Donard is well placed for Belfast and the Cliff House is equidistant from Cork and Waterford. The mixture of catchments is noted at Bellinter House:

> Most of our clients come from Dublin. Some more would come from the North (Northern Ireland) while others are more local and come here more for the restaurant food and for treatments … so there is a sort of weekend/short-break market from Dublin and a local food/beauty market as well. (Respondent B1)

Several of the spas, including Delphi, Monart and Farnham, noted these mixed local/regional/national catchments, though their locations in relation to the national road network played a part as well. Delphi is particularly remote, and visitors really do need to get away from it all just to get there, though this may arguably lead to a more focused and dedicated clientele in terms of the therapeutic experience of place. Here distance as time is as important as distance as geography and invokes a therapeutic notion of time stopped or interrupted. Finally the form of the Yauvana Spa at the Kingsley speaks to more experiential dimensions of clienteles and catchments. Users of the spa noted by Respondent KG1; '… would be mostly local from the Cork area, but some come from a little farther away because of the therapeutic side'. She noted that the clientele were much more informed about Ayurveda than expected and put this down to travel and global experiences which created more specific expectations and performances of healing in place.

In the imagery of the modern spa, the brochures for Inchydoney, Farnham and Monart all show a setting which is designed as a shorthand for exclusivity, luxury and quality for specific markets. Those clienteles can be identified in a number of forms. The national tourism agency, Fáilte Ireland, have classified in geo-demographic terms, the typical users of health/wellness facilities (Fáilte Ireland, 2007). These include; fun-seekers, occasional pamperers, relaxers, serenity seekers, beauty queens and help seekers. In settings like Bellinter, typical clienteles are professionals in the 25–40 age categories while couples are the most common group at Monart. The 'professionals' angle is also noted elsewhere as at Delphi and at Cliff House, where a slightly wider age range was identified. However all of the spas identified a more mixed clientele, with family groups and less affluent visitors also common, with the latter group often having their visits covered by a gift voucher or present. At Monart, the clienteles are similar and there is a corporate dimension as well, also noted at Cliff House, Bellinter and Slieve Donard. Apart from Temple, the most wellness-oriented site, there was a relative lack of the more focused 'serenity' and 'help seekers', noted in the classifications above, though again it was difficult to make a definitive statements on this given, for example the clienteles at the Yauvana Spa. There appeared to be a broad correlation between the more holistic user and the destination spa, with the 'occasional pamperer' or 'beauty queen' to be found more commonly at the hotel spas. Yet even here the complexity of uses at Monart, Farnham and Delphi meant that there was considerable intermixing of clientele, intent and utilization.

The extent to which the modern spa is an open and accessible place for all is contested around narratives of wealth, privacy and cost. To many, the modern spa is symptomatic of the worst excesses of the 'Celtic Tiger', bastions of selfishness and the spiritual home of conspicuous consumption. While cost and affordability are issues, automatic assumptions that modern spas are exclusive and, by association, excluding, must be considered more carefully (Conradson, 2007; Lea, 2008). Exclusiveness has a mixed role in the global identities of the modern spa and evidence from Ireland suggests this to be equally true. At Monart, exclusiveness is deliberately built into the image of place, given wider competition and markets; 'Well yes it is exclusive and deliberately so … we can't possibly sell the product as cheap … we need to deliver and to compare with a hotel' (Respondent M1). Other spas were adamant that they provided good value for money with Respondent B1 at Bellinter noting;

> In terms of expensive, well I don't think Bellinter is expensive for what it is … there was a huge decision to be made as to what price to set it at … to be honest we can't afford to be too exclusive as there's a lot of competition out there.. so no we're not really that exclusive. (Respondent B1)

At Kelly's the price was seen as part of a wider hotel package which was again considered to be good value for money. While there is no doubt that almost all the modern spas visited are by no means cheap, this must be placed against generally high hotel prices in Ireland and the relatively high prices of other forms of complementary and alternative medicine, which consumers seem relatively happy to pay (Heller, 2005). Indeed, private levels of spending on CAM globally point to an acceptance of out of pocket expenditures, even in countries such as the UK and Ireland where nominally at least, there are comprehensive national health systems (Wiles and Rosenberg, 2001).

In addition, the role of 'healing', while more central to the identities of the destination spa, has associations with class, income and education. In the classic wellness settings, whether at yoga or contemplative retreats and holistic centres, the majority of the users are relatively well-educated, health conscious and avowedly middle-class (Conradson, 2005b; Hoyez, 2007a; Lea, 2008). This applies also to wider images of the CAM consumer (Heller, 2005). Yet these essentialist images are breaking down slightly and at Kelly's the notion of a presence or availability of wellness treatment means that the hotel patrons have a choice or opportunity; 'So they're experiencing the thermal suite for the first time or they're taking treatments for the first time ... and certainly people come on spa breaks or just come for a break and want to have a go' (Respondent KY).

In considering the notion of 'health as gift', it is also the case that friends or family see the role of a spa as providing hard-working and deserving people with a break or are motivated by the impulse to take care of someone who won't look after themselves. At Kelly's identities of place as a site of memory were noted with the hotel being used as a place of gathering for families remembering a loved one and even as a form of psychological refuge in terms of personal stress or hardship. In arguing that the modern private spa is less excluding and exclusive than its image and narrative suggest, it may be seen as defending the indefensible. But this argument reflects an acceptance of experience and choice, as well as a natural distrust of assumed identities which are often reworked and renegotiated by individuals in their ownerships and explorations of personal health in place.

Finally, spatial performances at the modern spa can be seen in their health histories and performed mobilities (Hoyez, 2007a). In their older health histories, continuous expressions of health were apparent at a number of sites such as Kelly's location in the sea-bathing resort of Rosslare. At Bellinter, discursive and recursive elements of restoration and retreat are explicitly noted; 'Being originally designed (in the 1750s) as a country retreat, a rural getaway from the rigours of city life. We like to continue that tradition with our treatment rooms and swimming pools designed to refresh body and soul' (Bellinter House, 2008).

Similarly Monart's original site as a colonial estate owned by the Anglo-Irish was reflected in a conscious decision by the current owner to 're-hibernicise' that history by replacing colonial and introduced tree species such as monkey-puzzles with native forms. Along with the explicit use of Celtic motifs in the garden design, these spatial shapings spoke to a deliberate process of 'designed decolonization'. The inclusion of a labyrinth, an established spiritual symbol with links to both pagan and Christian practice also suggested older sacred-health links. How these local metaphors of place design fit with the simultaneous use of globalized practices of geomancy represents an interesting paradox. The location of new forms of globo-colonial health at the Kingsley hotel is particularly interesting given that the hotel and spa were built on top of the former site of the Lee Baths (see Box 6.2). These public open-air baths on the banks of the River Lee were an institution for Cork people, of both middle and working class hue, and this palimpsestic expression of complex local-global health identities and histories reflects other mobile meanings encountered in the book. Finally mobile performances are visible in the movements in and around the sites and between interior and exterior spaces. This was expressed in the long curved corridors of Monart and in the separation of the SeaSpa suite at Kelly's where one was required to leave the main hotel block to access the spa. Even within the thermal suites at both settings, different treatment rooms and restorative spaces and movements between them were part of the spatial performance of healing.

Economic Performances of Health: Ayurveda and Agelessness

As is already apparent, economic performances of health are central to the modern spa and reflect wider overlaps with geographies of tourism and wellness (Urry, 2002; Smith and Puczko, 2009). While the selling of health in place has been a constant theme across all of the sites, a pattern has emerged whereby entrepreneurship has become more and more central. The underlying nature of that entrepreneurship has however remained relatively constant, even if its forms and mediums have changed. Reputation, habitus and fashionability play as much a role in the cultural and economic production of the modern spa as in the original spa town (Porter, 1990; Kelly, 2009). Entrepreneurial alliances of landowners, hoteliers, speculative investors are also represented at watering-places across the centuries. While modern marketing represents a variation in mode and form, the essential methods and messages of promotion, competitive comparisons, advertising and even word-of-mouth, remain essentially the same since 1700 (Lee, 2004). The scales of development in the modern spa form a continuum from the small hotel with additional spa elements (small-risk) to the large-scale investment in destination spas such as Monart. These visible investments, and by association, risks, are in part responsible for shaping the place of health and how that health identity is in turn shaped by wider economic requirements. It may also explain the relative importance of social and cultural aspects and where health and healing sit within these wider narratives of commodified space (Conradson, 2007; Hoyez, 2007b).

In European towns with continuous spa identities like Spa or Bad Homburg, the commodification of that healing identity has been transferred and economically revitalized in place (Porter, 1990; Lee, 2004; Smith and Puczko, 2009). In addition, the place of health in such settings is more defined and 'accepted', given their mix of orthodox and CAM practices and narratives across time (van Tubergen and van der Linden, 2002). However, in Ireland and elsewhere, the creation of new 'landscapes of spa consumption' have come with a requirement for a modernist, even post-modernist re-invention of both setting and health identity with many attendant risks and pitfalls (Smith and Kelly, 2006). In particular the image of water as product, it's marketing within a wider promotion of holistic health and wellness, how wider health identities are consumed and the significance of ownership and agency, all feature in the economic production of the modern spa space (Smith and Kelly, 2006).

As noted previously, images of therapeutic place, often incorporating water, are used extensively in the commodification of the modern spa. Glossily illustrated brochures and web-sites contain explicit narratives that echo those images. Inchydoney is represented by aerial views of the hotel and its location on a broad beach, while many images of Slieve Donard place its grand Victorian exterior against a backdrop of the Royal Co. Down Golf Club links and the nearby Mourne Mountains (sweeping, of course, down to the sea). Interior imagery is heavily focused on those social narratives of luxury, relaxation and pampering that meet the expectations of the consumer. These imaginative versions of what Conradson terms an 'experiential economy', are central to the marketing of place and the positioning of treatment packages is a major part of that process (Conradson, 2007). While these are implicitly health-focused, and indeed use embodied imagery of 'hands-on' massage and physical spaces of relaxation, they also lean heavily on notions of self-gifting, indulgence and personal entitlement. Thus at Inchydoney, the brochure suggests that; 'Our menu of treatments is a pleasure for the weary body. From the soles of your feet to the top of your head, we invite you to embark on a journey to a sensory heaven' (Inchydoney Island Lodge and Spa, 2008).

In this vignette, implying both a sense of consumption (via the menu) and a modernist take on the health pilgrimage (journey to sensory heaven), embodied and affective themes are used, which are then backed-up in the treatment menus where specific 'Breakaway Day' packages include titles like, Revitalize, Rejuvenation and Relaxation. In addition, the implied journey towards a 'fountain of youth', reflect the repeated metaphors of water as both the giver and the preserver of eternal life.

More broadly, the role of wider media, in the forms of print newspapers and television, are also implicit in the spatial production of imaginative experiences at the modern spa. As is the case in many tourism sites, reviews are included in the media for both regulatory and wider consumerist purposes (Murphy, 2008). Reviews of spas like Monart, Temple and Bellinter appear in that bastion of consumerist cultural production, the 'Sunday Supplement', as well as on television and in women's and increasingly, men's magazines. In the case of Bellinter, this process can be explicitly seen in its adoption of a particular beauty/health product, the Voya range. Based on organic seaweed, the Voya products provide a link to sea-bathing as they were developed by the established owners of the Strandhill Seaweed Baths. Yet in an act of pure post-modernist commodification, the relationship between the producer (the traditional seaweed-bath) and the re-seller (the owners of Bellinter) was formed through a reality TV programme, 'The Mentor', in which the entrepreneurs who owned Bellinter were looking for competitors to come up with a new product. As a result the reputation of Voya (who competed in the programme) has been transformed via extensive media exposure, and their products are now sold at Cliff House, Delphi and Monart as well as Bellinter.

These multiple narratives of product at the modern spa also reflect a wider selling of healing and wellness. Those restorative and wider holistic meanings of health focus strongly on a narrative which speaks directly to a particular vision of CAM as consumer product (Doel and Segrott, 2003). While this is most strongly expressed at Temple, at other spas, those associations with healing and wellness often sit in a complex position between treatment and product. While this contradictory position can be applied to much of the business of health care it is particularly visible in the selling of the modern spa. Thus the seaweed baths at Bellinter are presented as both a form of healing product and cosmetic treatment for the skin. Even at the Yauvana Spa at the Kingsley, the specific forms of ayurvedic healing such as *Pizhichil* contain affective carative and cosmetic elements:

> Pizhichil … the Yauvana luxury massage, classified in India since time began as a 'royal' treatment. The unforgettable experience of a generous stream of warm, specially-prepared medicated oil applied to the body by two therapists using specific strokes in synchrony. The combination of oil and heat ensures that the medicinal oils are carried deep into the tissues giving benefits that strengthen the immune system, give relief from muscular tension and ease pains in the joints. It has enlivening and fortifying effects to combat long-term exhaustion with anti-ageing properties and special softening and nourishing effects on the skin. (Kingsley Hotel, 2009)

In such treatment narratives, wider invocations of beauty, agelessness and sensuality are combined with the physiological to create a 'special' experience for the user. Yet these simultaneously focus on another important aspect of the commodification of the modern spa, the USP, or unique selling point. Just as 'authentic ayurveda' is the USP of the Kingsley, so thalassotherapy functions in a similar way at Inchydoney or a deeper holistic health serves to identify unique performances of health at Temple.

In these processes of entrepreneurship and commodification, it is also worth asking where individual/communal performances and ownerships can be expressed. The spa and wellness business is well represented at all of the sites via menus of treatments and products. In this mix of stillness and commerce, trying to negotiate one's individual affective space can be a complex process. While for some, this is found in the hands-on massage or the seaweed wraps at Inchydoney, the difficulties of mixing this need for stillness and restoration in public spaces was noted in the observation by a user/reviewer that Saturday was; 'a very busy day with a lot of day spa guests crowding the room. This did not exactly provide the peace and tranquility the brochure promises and which you need after a 55-minute well-being massage' (Thesing, 2004).

While some of these tensions were worked out by making spas adult-only spaces, these exclusions were aimed at providing a setting and space where health and wellness dimensions of relaxation and stillness were more fully developed (Conradson, 2007). In some of the sites, such as Kelly's and Bellinter, the spa spaces were separated from the main hotel area, while even in the carefully-designed setting of Monart, the accommodation, public and thermal spaces are separate and distinct within the wider site.

Table 6.1 Fáilte Ireland Spa Classifications, 2008

Source: Fáilte Ireland, 2007.

Given that economic performances and competitive aspects of pricing and services remain important, issues of agency in the form of regulation and standards re-emerge. Fáilte Ireland has, belatedly in some views, come up with a classification system for the modern spa based on international standards. This has been developed in conjunction with spa owners, in part as a response to entrepreneurial concerns around the indiscriminate use of the word spa. Respondent M1 saw his destination spa as a specifically authentic form and bemoaned the lack of regulation, whereby any hotel could attach a small Jacuzzi or hot tub and call itself a spa. In answering his own question; 'When is a spa not a spa? When it's a hotel', a modernist notion of 'place quackery' was implied. A range of titles, such as destination spa, hotel (bijou) and day spa have therefore been developed, accompanied by short descriptions of the types of services and treatments expected in such settings (see Table 6.1). There are intended to inform and even shape consumer expectations of a restorative product which has some form of symbolic healing/well-being dimension. The importance of reputation and consumer confidence is also embedded in what users experience in the modern spa space and any sense of false advertising is closely monitored. There is an amount of self-policing here as noted by Respondent SI:

> There is also a lot of variety in how the spas themselves are classified. In terms of categories there are what I would call 'loose buckets' in terms of classification guidelines. These have been drawn from international/US models and we have tried to follow these in setting out the classifications on the site. Fáilte Ireland

have also got involved in conjunction with the owners of the spas but there is limited outside scrutiny on this – in a sense the spa managers/owners are self-policing in this regard.

The form of contestation of health identities within the spa and wellness business may also be telling. As visible landscapes of consumption, even conspicuous consumption in class identity terms, they also function within global commodity chains. As a result they may not be seen as a threat to orthodox medicine, both due to the (careful) lack of health claims by the providers, and the limited regard in which they are held by the medical profession (Stone and Katz, 2005). In this, the modern spa is less contested when compared to the older watering-places, which may reflect not only the entrenched power of the biomedical paradigm, but also the changing nature of capitalist society where health is seen increasingly as a business rather than a service (Kearns and Barnett, 1999).

Yet these experiential encounters are expressed communally, which assigns a stronger agency to the consumers of health in place. While political economy notions of structure and power are undoubtedly at play in the modern spa, the ephemerality and vulnerability of that power must also be seen in the light of the commercial power and habitus of the consumer, especially where the product is seen as a luxury spend. Fashionability and reputation, core concepts within therapeutic landscapes, may need to be seen in more explicitly economic terms of consumer confidence. As elsewhere, the extent to which the modern spa will sustain or be ephemeral (and some North American destinations spas date back to the 1940s) will be dependant on the interplay of commodificatory discourses and their specific impacts on the user (Frost, 2004; Cant, 2005). As an example of the shifts in response, Cliff House, Monart and a number of other spas are in discussion over the formation of a consortium ('the Inspire Group') where specific identities of stillness and quiet are expressed through their re-commodification as adult-only spas, from which large and potentially disruptive groups will be excluded, to create a properly restorative space (Conradson, 2007). This deliberate creation of geographies of tranquillity and stillness is intended to promote a specific type of restorative spa, aimed at ironing out wider tensions over other utilizations and at creating distinctive identities of healing and wellness. In the recent shift towards a greater appreciation of the place of health at the spa, as noted at Slieve Donard, this process may be already happening, though it may be of a more contingent and managed form. In trying to understand the psychological and affective performances of health at the modern spa, their conflicting cultural production and commodified identities always get in the way.

Summary: From Distress to Destress

> Be yourself again … escape the stresses of modern life and arrive to a safe haven, a nurturing environment where all that matters is you. An adults-only

environment ensures that privacy, peace and exclusivity are paramount. Relax, unwind and allow yourself to breathe, absorbing the healing effects of nature. A time for you to discover your inner-self. (Monart Spa, 2009)

In her work on community spaces of mental health, Vanessa Pinfold (2000) cited a respondent's search for a 'safe haven' within a chaotic external world. In seeing Monart as a safe space, one can also see the place of the modern spa as a retreat and refuge, where one ostensibly absorbs, 'the healing effects of nature'. While comparisons between a luxury spa and community psychiatric settings may appear specious, even insulting, there are connective threads in the processes around which such spaces are culturally constructed and psychologically performed to provide healing benefits. The notion of the safe haven, wherein one can withdraw from a stressful world, to a setting and time-space where a more personal inner journey can take place, also reflect wider constructions of health in a mind-body-spirit sense. While there are clearly genuine inequalities in the forms by which those external stresses are alleviated, the therapeutic settings of the modern spa, both natural and constructed, do offer places/spaces which can be accessed and used by different users for a range of different therapeutic experiences.

Those emotional, physical and belief aspects of healing and wellness also enable a wider vision wherein contested views of the modern spa as 'health site' can be challenged. That contestation is part of a wider debate around the value(s) of CAM and even to an extent, new forms of spirituality (Clarke, Doel and Segrott, 2004; Heller, 2005). While medical efficacy and evidence may be absent from a rationalist, modernist viewpoint, it is against a wider view of health and wellness, rather than physiological monitoring, that the benefits should be measured (Curtis, 2004). In truth, verifiable medical efficacy has, at all of the places in the book, been blurred and contested and all represent sites where the tensions and interplays of the biomedical versus social models of health are enacted. Selling the health-giving properties of place is always contested, though the mention of stress may help. Cancers, strokes and heart disease still account for around two-thirds of all deaths in Ireland and while there are multiple causal factors for all of these, stress and modes of modern living are increasingly accepted as significant factors (Pringle and Houghton, 2007). The modern spa, for all its pampered narratives, does still function as a site of restoration, stillness and wellness (Conradson, 2007; Williams, 2007).

In the material construction of the modern spa ensemble, therapeutic elements are both designed and implied in the built form and structural design. Water remains at the heart of therapeutic performance. In the external spaces, water is represented by the sea at settings such as Kelly's, Slieve Donard and the Cliff House while the natural landscapes of river and lake are associated with healing spaces at Delphi and Farnham. At Inchydoney, those specific therapeutic identities are expressed in the thalassotherapy treatments. Interior uses of water are built in to the designs and treatments to be experienced at Monart, Temple, Bellinter and The Kingsley. Materially there are connections to place histories where recurrent health identities are both re-commodified and altered. The Cliff House sits on a site close to St Declan's Well and uses the name, The Well, for its spa. Those spatial narratives of refuge and hermitage also spill from the well and hotel settings down the cliffs to the restorative sea below, as part of a centuries old maritime estate of healing which is simultaneously landscape-as-gaze and landscape-as-lived (Wylie, 2007). Native forms such as seaweed baths are co-located with the more exoticized treatments and spatial designs of ayurveda and the *hammam*. The vertical reproduction of bathing space at the Kingsley exemplifies this elision of vernacular form (the Lee Baths) by a reproduced global form (the Yauvana Spa). In these material tensions between local and global, and native and colonial forms, the material aspects of the hydrotherapeutic ensemble act as representative spaces of both health and cultural histories.

Those spatial dimensions of therapeutic landscapes are further enhanced in the cultural and commodified performances whereby health is enacted in modern spa spaces. There are particularly strong associations with spiritual and holistic forms of health, especially in the more focused destinations spas of Monart and Temple, where well-being and wellness represent an at times under-played carative and curative narrative. Indeed at Temple, Respondent TE noted an increasing interest in mindfulness-based therapies and even a potential expansion into holistic forms of hospice (echoing recent changes of meaning at the holy well). However, given the metaphoric nature of those narratives, modern spas are sites of continuous contestation, especially around CAM and beauty treatments and continue to be seen as problematic by orthodox medicine (Heller, 2005). Yet stubborn healing reputations, central to definitions of the therapeutic landscape, are sustained at the spa watering-place (Gesler, 1998; van Tubergen and van der Linden, 2002). All 10 of the sites studied either use the term spa in their full name or in an associated facility; discursive links to relaxation, cleanliness and de-stressing. Aspirational message of healing are implicit in the documentation produced by all of the sites. At the Kingsley the combined options of Ayurvedic spa, thermal suite and health club all invite the user to; 'Enjoy an authentic and all-natural holistic opportunity to gain greater health of mind, body and soul' (Kingsley Hotel, 2009). These links to a reputational and relational consumption of health and wellness rather than an evidence-based outcome continues to give the modern spa an imaginative health role.

In seeking to unpick some of the tensions inherent in those metaphors of place=health/wellness, useful ideas can be drawn from those theoretical underpinnings which relate to specifically embodied performances of health (Conradson, 2005b). In many of the spas, those embodied aspects of the therapeutic encounter or experience are represented in the treatments which link body to water and which echo the wider individual histories of human engagements with spa/bath/spring/pool. In the hydrotherapeutic encounter of the *hammam* or Kneipp walk at Monart, or the *Shirodhara* treatment at the Kingsley (where medicinal oils are poured on the forehead), embodied treatments inspire embodied and affective responses. In the performances of wrapping and immersion, either in seaweed as at Inchydoney or in seaweed baths at Kelly's or Bellinter, treatments are physically embodied. Mobile bodies are inherent to the modern spa, where silent, white-clad figures, de-gendered in identical white dressing gowns and slippers, move almost ghost-like from sites of treatment to sites of rest to sites of nutrition.

In containing inhabited dimensions, individual or group behaviours are essential parts of therapeutic enactment (Lea, 2008). Culturally produced and socially performed acts of healing and wellness, linked specifically to emotional and affective aspects of recovery and rejuvenation, are individually negotiated, despite a subtle regulation. That regulation of behaviour is at times obvious through specific rules and expectations or in the almost church-like atmosphere of the recovery rooms at Bellinter or Inchydoney. Indeed the atmosphere of silence at some spas seems almost unnatural in an Irish setting where conversation is central to cultural identity. Yet one's own feelings and even a form of liminality can be enacted in internal silent space, where expressions of geographies of stillness are given performative and personalized expression (Conradson, 2007). One can also see in the mobilities of the modern spa echoes of earlier performances in the watering place, wherein new privatized 'promenade spaces' are relocated from public to private settings. In such settings, and echoing the Turkish bath, the removal of outward badges of status may be seen as a form of social and cultural levelling (Breathnach, 2004). Finally, images of luxury, pampering and elite clienteles may not always be matched by inhabitations. Through the notions of gifting, competitive offers and indeed the economic realities of the global downturn, such exclusivity may be challenged. In the ability of the modern spa to survive a period where entrepreneurship is likely to be fully tested, the place and placing of health, in the buildings, discourses and behaviours at the modern spa, may repay closer attention.

Performances and experiences in place continue to connect the modern spa to its predecessors as representative and inhabited therapeutic landscapes (Conradson, 2005b; Williams, 2007). In such psycho-cultural understandings one can recognize that orthodox explanations around embodied landscapes of cure and repair are insufficient to explain a wider geography of mind-body-spirit wellness (Williams, 2007). Instead, modern spa settings must be seen in terms of a wider set of landscapes of alleviation, where care (and even self-care) is as important as cure and where a wider well-being is expressed through health management and maintenance. In the individual's decision to go away for a few days to Monart or Delphi, one can envisage a metaphoric health pilgrimage that is spatially enacted in a journey away from home/work and towards a more whole self (Smith and Kelly, 2006). This can be critiqued as an individual indulgence (given that a peaceful space can be accessed anywhere for free) or a choice only available to the affluent. In addition it can be argued that true serenity seekers are more likely to be found in ashrams, retreats and wellness centres than at spas (Lea, 2006). A more socialized model, as formerly practiced in continental Europe, could open up access and challenge current cultural constructions. While health performances may be ambiguous and difficult to measure, they reflect older histories of health in place where similar enactments and experiences gave rise to a variety of outcomes, both positive and negative. The retreat to the safe haven, where one can, through water, immerse oneself and allow wider stresses and cares to literally wash away, can be seen as a form of health-risk strategy. In the enactment of a carative shift from 'distress to destress', one can, especially in the more healing-focused setting of a destination spa, see a parallel to CAM responses to chronic and mental forms of illness. Finally, while the spatial form and design of the modern spa may differ somewhat from its eighteenth century predecessor, the wider performances of health and the self-negotiated therapeutic experiences remain robustly the same.

Chapter 7
Conclusion

Introduction

> Conceiving a notion of landscape as therapeutic is, itself, a construct, but one
> that is derived from a rich and complex set of individual and cultural meanings
> – many of which seem to originate from our earliest and most primal multi-
> sensory experiences. (Milligan and Bingley, 2007, 802)

In considering this interplay of individual and cultural meanings, there are strong
overlaps with Gesler's definition of the therapeutic landscape as containing aspects
identified with inner/meaning and outer/context (Gesler, 1993). In considering
individual/communal health performances, their spatial dimensions reflected the
formative natural settings and built environments central to Gesler's understanding.
Yet the other dimensions, linked in particular to symbolic meanings, sense of place
and everyday activities, connected more directly to a deeply embodied performance
of health. In tying this to Milligan and Bingley's notion of 'primal multi-sensory
experiences', these connections between the spaces and bodies of performance
become clearer. In theoretical terms the connections to a cultural landscape
shaped by 'maps of meaning', further strengthens these inner performative and
experiential elements of body and place (Jackson, 1993). Finally the connection to
Eyles understanding of place and well-being as expressed through daily experience
and perception's on one's place-in-the-world further emphasize the power of that
inner meaning (Eyles, 1985; DeMiglio and Williams, 2008).

Yet these individual/communal performances were also culturally produced
and confirmed Porter's suggestion that;

> They demonstrate the degree to which past cultures of health were complex
> performances – enterprises shared between the sick and the medical profession
> (itself intricately stratified); operating within a matrix of resources, institutions,
> amenities and physical buildings; and drawing upon elaborate rituals of regimen
> … They lasted ... because they satisfied a deep desire that the healing enterprise
> should proceed within frameworks essentially sociable in their nature, and
> suffused with symbolic cultural meanings. (Porter, 1990, xii)

It then becomes possible to see other performances of health, especially those
more closely associated with cultures and economies, as reflective of Gesler's
notion of an outer/context. In the beliefs and philosophies, social relations and
territoriality that made up that context, it was possible to see how a complex stew

of wider agencies, linked to power, profit and ownership, were also embedded and active in the watering place. Such cultural landscapes were also associated with a theoretical position which argued for the importance of social relations and a wider politicized shaping of performance motivated by moral values and financial profit (Mitchell, 2000). The lasting performative frameworks noted by Porter referred to historic spas, but could just as easily be applied to modern spas and older forms such as holy wells and seaweed baths. These tensions between affective experience and cultural production emerge from the book and the evidence and arguments contained within it, to suggest that what we witness at the watering place acts as a microcosm of wider health/place debates. In all of these tensions, negotiations and enactments, water provides a valuable metaphor/setting for this process.

Performances of Health

Water was at the core of these performances and the ways in which health and wellness were enacted, created, understood and contested were reflective of mobile narratives of water as an ambiguous healing product (Strang, 2004). That product flowed in turn from the cultural, spatial and economic structures and agents that shaped a wider performative meaning (Wylie, 2007). Yet it was also an embodied process which talked to an experiential engagement with place, wherein the 'relational self' and lay therapeutic meanings were negotiated and owned (Conradson, 2005b). The book's focus on water and health was also set within a CAM understanding of healing and well-being rather than biomedical health. In part this was guided by the particular forms of practice engaged with at the different sites, where more complementary and alternative enactments produced a type of 'spatial wellness'. But it was also a focus linked to the construction of a landscape of therapeutic potential where self and communal place-health experiences were enacted in and around water. In overall terms, these types of cure at the watering place could be considered, in their generally restful response to a set of chronic and recurrent illnesses, as a form of 'slow health'. In this term, more holistic aspects of prevention, maintenance and restoration of health acted as a balance to the treatment of illness, so that they were as much spaces of care as spaces of cure. In its natural, hydrologically-assisted recovery, the body healed itself in its own time, an outcome which medication may have speeded up but did not fundamentally alter.

The types of conditions and cures encountered in the different watering-places show a surprising consistency across time. Even if not always fitting with biomedical evidence (though there were plentiful examples, especially when considered against then-current medical knowledge), the accounts of cures in place reflected a mixed natural/cultural understanding of the healing power of water and a wide range of therapeutic encounters and experiences (Logan, 1972). While some of the panaceal associations, especially at the holy well and spa town, spoke to an aspirational and hard-to-credit power in the water, the main types of cures could be broken down into two broad groupings, chronic physical conditions and mental health. While few would make any claim to the power of water in acute medical emergencies, apart from severe dehydration, healing was strongly associated with chronic and recurrent conditions. Skin cures, both cosmetic and chronic, especially for conditions such as eczema and psoriasis, were associated with holy wells, historic spas, sweat-houses, Turkish baths, seaweed baths and even the modern spa (Varner, 2002; Kelly, 2009). The presence of patients suffering from a range of chronically debilitating conditions at most of the sites confirmed the assumed benefits to be gleaned from immersion in, and consumption of, the water. These included, digestive (ulcers and liver), respiratory (asthma and emphysema) and mobility (arthritis, rheumatism and feet) conditions. Other chronic illnesses fell into otorhinolaryngology (eye and throat), dental and genito-urinary (gonorrhoea, fertility, urinary) categories. Simple fevers were treated at spa towns and the two sweating-places for which purposes it was especially effective.

A second set of cures, tied up as much with the place as the water, were those which related to mental health in its widest sense (Heller, 2005c). The therapeutic power of place was frequently mentioned and was reflected in the natural settings, layouts and therapeutically designed spaces of holy wells, spa towns, sea-resorts and modern spas (Williams, 1999c). From simple cures for headaches associated with a range of wells and the modern spa, to the traditional complaint of nervous disorders, generally treated at the well and Turkish bath, the soothing and restful impact of the watering-place had a solid reputational history (Smith and Kelly, 2006). In wider expressions of recovery, rest and recuperation, the modern spa, spa town, well and seaside settings all represented the healing power of water in a distinctly affective form, where embodied notions of de-stressing and detoxification represented the removal of illness as well as the restoration of a form of balance within a poisoned mind or spirit. In the journey from distress to de-stress, the notion of the rest cure was expressed at a number of sites in more specifically psychotherapeutic ways (Conradson, 2005a). The use of St. Ann's as a sanatorium where shell-shocked and mutilated soldiers were treated during World War I, suggests a combined physical and mental curative space (Durie, 2006).

While these positive aspects of health were visible in all of the sites, they all had potentially anti-health dimensions as well. In many cases these was expressed not in the water itself but through the more liminal behaviours at the watering-place. Excess in the form of over-consumption and over-socialization meant that for many, the visit to the spa, well or sea-resort meant one came back as ill as one went. In the wildness of the pattern, the gambling-rooms of the spa or even the promiscuous beach, the social behaviours threatened health, from drink, gluttony, sex and violence. Ironically, it was in the supposedly most promiscuous and open society, the present day, that these behaviours were least expressed, if the modern spa can be used as a yardstick. It may reflect the specific class identities associated with the spaces, but has strong connections to prevention, regulation, risk and the development of wider cultures of health and safety. While an awareness of danger at the traditional setting of the spa, sweat-house or Turkish bath was visible in the fact that patients with certain conditions were not allowed to partake of the cure, there was a greater freedom around behaviours in such settings.

In seeing health as a performance or practice, there were connections to medicine's own terminology, but in its embodied form, that practice had specific experiential and affective meanings (Thrift, 2004). In the consumption of often unpleasant well, spa or sea-waters, or in the immersion in a slimy seaweed or mud bath, the direct bodily engagement with water was part and parcel of the curative performance. Yet there were ambiguities and contestations in that performance. For the sick farmer on a Leitrim hillside, the use of a sweat-house was simply a matter of pragmatism and choice; a performance of health to assist in the performance of labour. For the Anglo-Irish lady or gentleman, the treatments to a body enlarged by rich dietary habits and tastes, were a consequence and simultaneous reflection of an embodied performance of leisured identity. For the modern spiritual pilgrim to the well or spa, that performance is still an embodied process, albeit one where the diseased body is less a part of that performance as a diseased or troubled mind or spirit (Shaw and Francis, 2008). Yet all are in their essence somatic, and in that view of the body as site of health, the interactions of bodies and water in place is common across all of the examined settings in time and space.

In considering those more performative, experiential and affective aspects of the water-cure, psychological and wider psycho-cultural dimensions of place can be clearly identified. There are connections to recent work on self-landscape interactions and the shifting understandings of a therapeutic landscape experience, though in this case, these are set within more traditional settings (Conradson, 2005a; Lea, 2008; Dunkley, 2009). In addition, wider 'performative turn' concerns are enhanced by using the watering-place as a setting within to begin to examine affect, emotion and the lived mobilities of the body, mind and spirit in place. While these were 'owned' as experiential responses to one's own health in place, they were also performed, thereby providing an experiential production of health. Such productions should not be seen solely as autonomous and selfish acts, detached from any wider cultural meaning or structured shaping, but rather reflective of a link between the personal and the political, the psycho- and the cultural. Reflecting recent theoretical concerns with landscapes of therapeutic experience, personal engagements with place often had affective dimensions (Crouch, 2000; Conradson, 2005a). For some, the experience of a dark well, a claustrophobic spa town assembly or the sweaty and suffocating confines of the *sudatorium* could be profoundly anti-therapeutic. For others floating in an outdoor sea-bath, strolling along a spa walk or sitting in a modern spa recovery space, staring at water, can all be seen as positive and relaxing experiences from which people derived both physical and mental benefits. While these experiences are more heterogeneous than homogenous, one can see a positive potential, which, though not applicable to all, functions for many.

In the tropes of health at the watering place, the ongoing cultural productions and performances spoke as much to the social as the explicitly curative. Over time the metaphoric dimensions of the water-cure, from the spiritual healing at the holy well to the more medicalized baths at Blarney were bound up in a culturally-shaped engagement with water. In both the traditional visitor to the holy well or the Anglo-Irish patronage of the sea-bathing resort, one could see a range of health-place identities being expressed and enacted. Through shifting relationships between the water-medicine as pill, and its social performance, as sugar, those metaphors of practice stretched and modified the health identities of the watering-place. At settings like the holy well, the metaphorical health meanings of the votive/left offerings encompassed the entire life course from early life and even pre-life to death and became, for users, places that were representative of everything from site of 'surrendered symptom' to site of memory and loss. Yet for all that the cultural production of the healing-water place began to explain the identities of those places, it could not provide a full understanding of those identities. People engaged with healing places on their own terms, drawing their own meanings, values and individual interpretations of the body, spirituality and diverse understandings of health.

In considering the concept of 'being-in-place', encounters at the watering-place were reflective of individual mobilities and wider health paradigms (Lorimer, 2008; Thrift, 2008). Phenomenological encounters happen wherever people meet and the mobilities that occurred in and around the spas, wells and baths exemplified this. The notion of a health pilgrimage can be applied to all of the sites, in that there is a sense of a journey 'towards' communal encounter and the potential of a therapeutic positive experience at the site. This can be seen as self-affirming, and the consumption of health in the watering-place was connected to identity formation, especially for the Anglo-Irish (but also the Celtic-Tiger Irish). Yet, there was also a therapeutic value in the journey itself, a factor often associated with a sort of personal search for the other, or in Jungian terms, one's own shadow, to provide a mental component to parallel the physical re-balancing at the spa or well (Philo and Parr, 2003). Cultural performances of a more liminal kind, reflecting the social aspects of the watering-place were also, it cannot be denied, an attraction. In boisterous, gregarious, clandestine, celebratory and luxurious encounters in place, social mobilities were experienced and class identities confirmed and occasionally overturned, especially in matrimonial and other outcomes.

In the physical settings, the natural and built elements were visible expressions of a 'medical topography' or a spatial production of wellness. In the solid underpinnings of stone at the holy well, sweat-house, assembly room or Turkish bath, the healing waters were stored, enclosed and contained to be accessed by producer and consumer alike. Those same producers and consumers represented a range of people; material embodiments of the vernacular, the colonial, the exotic and the valetudinarian. In the changing material forms, shifting allegiances and expressions of health paradigms were also visible. As the holy well changed from a natural to a more catholicized form, so too the traditional exotic Turkish bath gave way to the more medicalized hydropathic treatment rooms where electricity and white coats replaced the Ottomanic colours and attendants. Specifically therapeutic aspects of place, associated with remoteness and silence were found at St Kieran's Well or in the remains of a sweat-house on a Cavan hillside. Such natural therapeutic and affective elements of place were paralleled by the more constructed and deliberately designed forms as in the Turkish Baths of Lincoln Place or the Spa House at Mallow, both discursive creations of colonial and Orientalist narratives of health imported into an as-yet colonial setting (Breathnach, 2004).

In their mobile spatial and curative identities and utilizations, many of the watering-places discussed in the book could be seen as ephemeral. In seeing ephemerality, not as a pejorative term, but rather a realistic representation of a mobile form, water's fluid engagements with health and place were evident. In addition, ephemeral experiences could be unconscious and accessed through affective and therapeutic memories of place. At the watering-place, an assumption that experiences, enactments and encounters were fixed and structured was also hard to validate. Some of the watering-places themselves were short-lived, such as individual Turkish-baths or spa towns, but the latter form lasted for almost 300 years. Even the most ephemeral of the forms, the Turkish-baths, lasted for over one 100 years. At the other end of the scale, holy wells dated back into the first millennium yet remain active therapeutic settings today, while the modern spa incorporates curative elements with Roman and Sanskrit antecedents. A recurrent ephemerality can also be seen in the personal uses of the watering-places, wherein habit and inhabitation met. While for some people, visits were one-off events, for many others the practice of the visit to the watering-place was habitual, recurrent and a repetitive encounter. Some of this habitual use was tied in with the seasonal performances of the pattern or the summer rest-cure. In their common identities as sites of chronic cure and illness, the utilizations of the watering-place also reflected a recurrent health need. In sometimes dismissing the watering-place as a setting where the instant cure could not be found, one can ignore the fact that within much of current western biomedicine, absolute cures, for both acute and chronic conditions, remain equally stubbornly elusive (Heller et al., 2005).

Finally, one can see the performances of health in these places as worked out under a range of 'managerial' gazes which, due to the underlying liminality of almost all the places, created space for contestation and diverse interpretation. A whole series of culturally-formed 'gazes', medical, clerical and colonial, shaped and controlled curative metaphors and meanings. In considering the ways in which experiential aspects of wellness were culturally managed and regulated, the notion of ownership is a valuable one (Ashworth, 1994; Smith and Puczko, 2009). In considering ownership, it can be seen in three different ways, drawn from three different constituents of the therapeutic landscape, place, culture and people. In place terms, that ownership included the actual land, the surrounding setting as well as the properties and ancillary elements. Just as the fields in which holy wells and sweat-houses stood had owners so more visibly did the Turkish baths and modern spas. In economic terms, the water and its attendant forms were physically owned and managed. While this ownership was often communal, such forms of ownership were mostly found around the vernacular sites. In the wider settings of the colonial watering-place, ownership was associated with historical patterns of power, structure and agency via the Anglo-Irish elite and the monied classes and many of the spaces were private rather than public.

A second form of ownership was visible in the cultural narratives of health in place. Confirming similar associations found in almost all therapeutic landscapes, the ownership of curative reputation was passed down and indeed carefully guarded by a range of producers and consumers (Kearns and Gesler, 1998). In the *dinnseanchas* of place lore and collective memory, the vernacular cultural creations of 'the cure' at the well, sweat-house or seaweed bath survived and thrived on metaphor and the specific cures and curees who used those places. One can even envisage these as oral performances of health. In the more constructed colonial and neo-colonial forms, those narratives were expressed in a colonial vision of a place-based 'sick trade' as well as in the exoticized narratives and practices of the Turkish bath or Ayurvedic spa. Finally in terms of a personal ownership, those narratives sometimes clashed and reflected deeper divisions in how people approached and managed health, so that the places became, if not quite a battle-ground, then a site within which wider metaphors of health, belief, faith, science and modernity were enacted and discussed (Williams, 1999a). It may even be that through those metaphorical discussions, deeper political, spiritual and philosophical notions of order and chaos were also enacted in a spatial form.

In the exploration of holy wells, spas ancient and modern and the spaces of the sea and sweat-cure, a number of useful connections between contemporary theory and traditional setting were established. In identifying that the therapeutic landscape experience was primarily affected by its performance, the healthy and sometimes unhealthy nature of that experience was in turn shaped by the construction of that enactment (Conradson, 2007). In the bodies, cultures, spaces and economies of those performances, the role of water as a curative form was both consistent and persistent. In considering the notion of globalizing therapeutic landscapes, the examples uncovered in Ireland suggested that this should also be understood in relation to local manifestations. Locally produced and directed forms such as seaweed-baths and sweat-houses marked variants from the global forms yet emphasized Strang's contention that, 'like individuals, societies do not discard earlier models so much as build on them epigenetically, so that the future is always a discussion of the past' (Strang, 2004).

In investigating these links within the watering-place where paradigms of health are fluid, cloudy and polysemic, I return to the word impatiences. While this may seem an odd word to use, to me it describes perfectly the clash of rational and irrational responses to health in these landscapes. In particular, biomedical concerns with truth, authenticity and proper scientific evidence are often expressed as an impatience with the superstitious, irrational and quackery-prone world of the hydropathic 'pseudo-cure'. Yet what is impatience but an emotional response. In the very act of explaining ones beliefs, one cannot but betray a belief-based positionality. It is precisely this ability of the ambiguous water-cure to foster confusion, contestation and impatience which was a motivation to work in this area (Strang, 2004). Though I am quite capable of a personal impatience with some aspects of complementary and alternative health, and in particular their exclusive and at times privileged manifestation, their hydrological expression provides valuable settings within which to explore truths and authenticities across health paradigms as a whole.

Cross-cutting Themes and Future Directions

This book has consciously sought to take a broad approach to the subject in an attempt to uncover the rich potential still to be found in traditional therapeutic landscapes. While there is a strong historical dimension to the work, it does reflect contemporary practices and settings as well. While the main themes of the book have focused on performances of health, a number of valuable cross-cutting themes have also emerged, some of which are noted here for future development and discussion.

In health terms, the life-course has been mentioned at different times in the book. The rich hydrological metaphors of the water-course could be explored more fully to trace the connections with the life-course, especially in its health dimensions (Strang, 2004). From the use of wells and spas (to aid the production of life), to the maintenance of healthy living and well-being (from recurrent visits) and on to the ways in which therapeutic landscapes can aid after-life healing (in the processing of death, grief and memory), metaphors of water-health-life have rich potential. In passing comments on gender, a number of spatial separations and social practices were observed, which identified many of the watering-places to be both gendered in their construction and also in their inhabitations and performances. While cultural dimensions of women's health appear in some of the spa narratives, the extent to which many of the sites are strongly gendered (especially in the modern spa and Turkish bath) bears further scrutiny. Loosely connected to gender, in its historic representations of sites of male-female (and indeed, same-sex) interactions, the notion of liminality has been dealt with more fully in the text, yet there is considerable potential in developing the work of social historians into this broader configuration of watering-places. The original violence of the holy well pattern or the wild behaviours of the spa-town or sea-resort were one form of this liminality, with a clear ability to become an anti-therapeutic performance or experience. From the more genteel matchmaking at Lisdoonvarna or Mallow, there is in addition, some definite potential in a fuller uncovering of the place of sex/romance, both in heterosexual and less documented homosexual terms at the watering-place.

From a spatial/cultural perspective, the role of colonialism, though covered in different settings in the book, could be more fully developed. In uncovering the role of colonialism in the production of Irish spaces and places, there is much that could be applied to other settings world-wide, especially in the inhabitations and performances of colonial identities (Kavita, 2002). This may even be applied to a notional 'colonial health', and again a deeper comparison between the role of colonialism in shaping attitudes and utilizations of health has value, especially when compared with 'native' performances of health such as the holy well or sweat-house. The arrival of exotic forms such as the Turkish bath (ironically re-exported from Ireland in Barter's commercially successful Victorian franchise), pointed the way towards a more contemporary presence of similarly globalized and proto-colonial forms of ayurveda and acupuncture.

In returning to a theme noted in the introduction, moral geographies deserve a final mention. The watering-places discussed in this book appear to cause strong and often divided opinions in people. Yet the watering-places persist(ed) and thrive(ed) in a variety of forms which point to a strong current of affective and material production (Mackaman, 1998). In their often ambiguous productions the watering-places challenged orthodox practices of the day, yet also represented them. In their new or dissident positionalities they seemed to bring out the judgemental in people. Those judgements were often couched as strong moral positionalities such as right and wrong, correctness and falseness, authentic and pseudo, healthy and unhealthy, elixir and poison, sacred and profane. Yet it is not the intent of this work to sustain the widely used, but often unsatisfactory metaphors of duality and polar divides. Rather, the different positions around which people discussed and argued the merits of the watering-place should be seen in fluid terms, mirroring for example, a Celtic knot or the linear shape of a river or lake and its sometimes complex channels and branches. In addition, the consideration of many of the sites as 'faithscapes', points to an additional moral geographical connection (Taylor, 1995; Shackley, 2001). While the holy well and modern spa are the settings most connected to old/new spiritualities, the power of the healing waters at all spaces was essentially founded on faith and belief in the material and metaphorical cures inherent in them. In the inhabited and performed aspects of person, health and water, those more subjective and experiential elements of the faith cure were realized as occasional manifestations of the sacred (Eliade, 1961).

More specifically, and this is an omission I acknowledge, a greater focus on user perspectives would be a valuable extension to this initial foray into the spaces of water and health (Lea, 2008; Lorimer, 2008). Constrained by the primarily historic focus and the private setting of the more contemporary work, the user voice is nothing like as strong as it could be. In part this is due to the sourcist nature of historical material, but some evidence such as the National Archives material from the 1930s, as well as local histories, points to ways in which oral experiential testimonies have been saved and stored. Organizing a fuller research agenda around affective responses to the watering-place, either through more structured qualitative research in a contemporary setting, or as a piece of archival work, should be possible, if methodologically complex. In carrying out such research at a global level, some valuable connections to how water, health and place vary in a cultural sense should also emerge. It may be that deep down, just as with the multiple global forms of the sweat cure, the same broad messages of spiritual, chronic yet acculturated health performances may emerge to become a new strand in therapeutic landscapes research (Aaland, 1978; Williams, 2007).

Glossary

Ague	A fit of fever or shivering or shaking chills, accompanied by malaise and pains in the bones and joints.
Ayurveda	A traditional Hindu system of medicine practiced in India since the first Century. Literally means 'knowledge or science of life'.
Balneotherapy	The treatment of disease by bathing. The term is generally applied to everything relating to spa treatment.
Bania	The Russian form of a sauna. Bania (or Banya) buildings can be quite large with a number of different bathing areas or simple wooden cabins.
Cachexies	A traditional name for general ill health with emaciation, usually occurring in association with cancer or a chronic infectious disease.
Chalybeate	A general name for a mineral spring or source impregnated with or containing salts of iron.
Chlorosis	A form of chronic anemia, characterized by a greenish-yellow discoloration of the skin and usually associated with deficiency in iron and protein.
Chthonic	Referring or relating to the underworld.
Currach	A type of small boat with a wicker frame and covered in skin and tar. Still used along the West coast of Ireland.
Dinnseanchas	A type of 'place lore'; poems and tales which relate the original meanings of place names and constitute a form of mythological etymology.
Dosha	The term used in ayurvedic medicine to classify patients by their mind and body type. There are three core doshas, *vata*, *pitta* and *kapha*.
Douches	Treatments that clean part of the body or a body cavity with a jet of water or air.
Dropsy	An old term for the swelling of soft tissues due to the accumulation of excess water, now referred to as an edema.
Garland Sunday	The last Sunday in July, a common date for celebrations in Ireland as the nearest Sunday to the old pagan festival of *Lúnasa*, the 1 August.
Hammam	A communal bathhouse, usually with separate baths for men and women, found across the Middle East.
Heliotherapy	Medical therapy involving exposure to sunlight.

Hurling	A field-sport played in Ireland with ash sticks and a leather *sliotar* (ball). A cross between field hockey and lacrosse.
Hypocausts	In ancient Rome, a system of central heating in which hot air from an underground furnace circulated beneath floors and between double walls.
Iatrogenesis	Illness induced in a patient by a physician's activity, manner, or therapy. Used especially of an infection or other complication of treatment.
Intendant	The name given to a powerful state-appointed official, usually a doctor, who had the charge of and managed spas in eighteenth- and nineteenth-century France.
Kermesse	In Europe, particularly in Belgium and Holland, an outdoor festival and fair.
Kneipp walk	A type of shallow cobbled water bath, usually for the feet and legs, called after its originator, Sebastian Kneipp (1821–97), a German priest and early proponent of hydrotherapy.
Kumbha Mela	A mass Hindu pilgrimage, which occurs four times every twelve years and rotates among four different locations.
Laconium	A dry heat treatment room or sauna which lets the body heat up slowly and gently, generally at a temperature of around 140°F.
Muezzin	The official of a mosque who calls the faithful to prayer five times a day from the minaret.
Nauheim baths	A bath in which carbon dioxide was passed through saline water to stimulate the skin and improve blood circulation and pressure.
Needle baths	More properly a needle shower, a type of treatment where jets of water were sprayed on the while the body from a circular frame.
Pardon	Patron saint's feast days in Brittany, often accompanied by small village fairs.
Penal Laws	A series of laws enacted in Ireland from the end of the seventeenth century aimed at the subjugation of religious, social and economic rights for Roman Catholics and Non-Conformists.
Pharmacon	A medicine or drug which can also be a poison.
Pisreog	A superstitious practice, charm or spell. Also spelt as *piseog* or *pishrogue*.
Priessnitz wraps	A traditional hydrotherapy treatment which involved a patient being wrapped tightly in wet (usually cold) blankets for at least an hour.
Quacks	A fraudulent or ignorant pretender to medical skill.
Rasul	An ancient Arabian bathing ritual combining a mud pack with an herbal steam bath.

Refridarium	A slightly tepid preparation room used in Turkish bath, where users changed into robes and wooden clogs.
Reiki	A system of hands-on touching by an experienced practitioner to produce beneficial effects by strengthening and normalizing certain vital energy fields within the body.
Russian baths	Another name for the bania (see above) or Russian steam bath.
Schnee baths	A form of galvanic electrical bath which used electric current and water. The typical four-cell Schnee bath had four different small baths for each limb.
Scrofula	Any of a variety of skin diseases; in particular, a form of tuberculosis, affecting the lymph nodes of the neck. Historically known as the 'King's Evil'.
Stone therapies	A form of body massage using smooth heated stones.
Sudatorium	A hot room in Turkish baths and other forms of sweating-places designed to make patients sweat profusely.
Sweat-lodge	A dome-shaped tent constructed by First Nations and American-Indian peoples in North America, used for ritual cleansing and purification by means of steam produced by pouring water over hot stones.
Temazcal(li)	A type of sauna native to Mexico, used for healing purposes. It normally resembles a small stone igloo with a low entrance and is made of adobe or stone.
Thalassotherapy	A general term for the medical use of seawater in a range of different forms such as bathing, seaweed baths or algal wraps.
Thermae	Roman buildings housing baths, though more properly applied to buildings with warm springs, or baths of warm water.
Valetudinarian	A sickly or weak person, especially one who is constantly and morbidly concerned with his or her health.
Wakes	More correctly termed 'wakes weeks', these were original religious festivals which became de facto workers holidays during the Industrial Revolution in England and Scotland.

Bibliography

Aaland, M. (1978), *Sweat* (Santa Barbara: Capra Press).

Agnew, F. (2003), 'The men who worked on the Dublin to Kingstown Railway', *Dun Laoghaire Journal* 12: 59-62.

Aitchison, C., Macleod, N. and Shaw, S. (2000), *Leisure and Tourism Landscapes: Social and Cultural Geographies.* (London: Routledge).

Anderson, K. and Smith, S. (2001), 'Editorial: Emotional geographies', *Transactions of the Institute of British Geographers* 26: 7-10.

Andrews, G. and Holmes, D. (2007), 'Gay Bathhouses: The transgression of health in therapeutic places', in A. Williams (ed.) *Therapeutic Landscapes* (Aldershot: Ashgate) pp. 221-232.

Andrews, G. and Kearns, R. (2005), 'Everyday health histories and the making of place: The case of an English coastal town' *Health & Place* 60: 2697-2713.

Andrews, G., Sudwell, A. and Sparks, M. (2005), 'Towards a geography of fitness: An ethnographic case study of the gym in British bodybuilding culture', *Social Science and Medicine* 60: 877-891.

Appadurai, A. (1996), *Modernity at Large: Cultural Dimensions of Globalisation* (Minneapolis: University of Minnesota Press).

Ashworth, G. J. (1994), 'From history to heritage – From heritage to identity; in search of concepts and models', in G. J. Ashworth and P. J. Larkham (eds) *Building a New Heritage: Tourism, Culture and Identity in the New Europe* (London: Routledge) pp. 13-30.

Atkinson, D., Jackson, P., Sibley, D. and Washbourne, M. (2005), *Cultural Geography. A Critical Dictionary of Key Concepts* (London: I.B. Tauris).

Austen, J. (1993) [1818], *Persuasion* (Ware: Wordsworth Classics).

Bakhtin, M. M. (1984), *Rabelais and his World* (Bloomington: Indiana University Press).

Barker, P. (1996), *Regeneration* (London: Viking).

Barter, R. (1861), *Descriptive Notice of the Rise and Progress of the Irish Graffenberg, St. Ann's Hill. Blarney* (Cork: St. Ann's Hydro).

Bassett, G. (1886), *The Book of Antrim* (Dublin: Sealy, Bryers and Walker).

Bayoumi, M. and Rubin, A. (eds) (2000), *The Edward Said Reader* (New York: Vintage Books).

Beirne, P. (2006), 'The Ennis Turkish Baths, 1869-1878', *The Other Clare* 32: 12-17.

Belhassen, Y., Caton, K. and Stewart, W. (2008), 'The search for authenticity in the tourism experience', *Annals of Tourism Research* 35:3, 668-689.

Bellinter House (2008), *Bellinter House Bathhouse Spa Brochure* (Bellinter: Bellinter House).

Berg, L. and Kearns, R. (1996), 'Naming as norming? 'Race', gender and the identity politics of naming places in Aotearoa/New Zealand', *Environment and Planning D* 14:1, 99-122.

Berger, L. and Rounds, J. E. (1998), 'Sweat lodges: A medical view', *The Indian Health Service Primary Care Provider* 23:6, 69-75.

Bergholdt, K. (2008), *Wellbeing: A Cultural History of Healthy Living* (Cambridge: Polity).

Berry, H. F. (1892), 'The Irish bath', *Journal of the Cork Historical and Archaeological Society* 1:6, 111-116.

Blaxter, M. (2004), *Health* (Cambridge: Polity).

Boorstin, D. (1964), *The Image: A Guide to Pseudo-Events in America* (New York: Harper).

Bord, J. and Bord, C. (1985), *Sacred Waters: Holy Wells and Water Lore in Britain and Ireland* (London: Granada).

Borsay, P. (2000), 'Health and leisure resorts, 1700-1840', in P. Clark (ed.) *The Cambridge Urban History of Britain, Volume 2* (Cambridge: Cambridge University Press) pp. 775-803.

Borsay, P. (2006), 'New approaches to social history. Myth, memory and place; Monmouth and Bath, 1750-1900', *Journal of Social History* 39:3, 867-889.

Bourdieu, P. (1977), *Outline of a Theory of Practice* (New York: Cambridge University Press).

Bourke, A. (2001), 'Introduction', in A. Rackard and L. O'Callaghan (eds) *FishStoneWater. Holy Wells of Ireland* (Cork: Atrium) pp. 7-12.

Bowen, E. (1942), *Bowen's Court* (London: Longmans).

Boyd, J. (1891), *Boyd's Pictorial Guide to Larne and the Antrim Coast* (Belfast: William Mullan and Sons).

Boydell, B. (1984), *'Impressions of Dublin – 1934'*, *Dublin Historical Record* 37: 88-102.

Boylan, E. (2002), *Tobernalt Holy Well: History and Heritage* (Sligo: St. Johns, Carraroe).

Breathnach, T. (2004), 'For health and pleasure: The Turkish Bath in Victorian Ireland', *Victorian Literature and Culture* 32:1, 159-175.

Bremer, T. S. (2006), 'Sacred spaces and tourist places', in D. J. Timothy and D. H. Olsen (eds) *Tourism, Religion and Spiritual Journey* (Abingdon: Routledge) pp. 25-35.

Brenneman, W. and Brenneman, M. (1995), *Crossing the Circle at the Holy Wells of Ireland* (Charlottesville: University Press of Virginia).

Brockliss, L. W. B. (1990), 'The development of the Spa in seventeenth-century France', in R. Porter (ed.) *Medical History, Supplement No. 10.* (London: Wellcome Institute for the History of Medicine) pp. 23-47.

Broderick, E. (1998), 'Devotions at holy wells: An aspect of popular religion in the diocese of Waterford and Lismore before the famine', *Decies: Journal of the Waterford Archaeological & Historical Society* 53: 53-74.

Buckley, A. D. (1980), 'Unofficial Healing in Ulster', *Ulster Folklife* 26: 15-34.

Buckley, J. (1913), 'An ancient hot-air bath-house, near Schull. Co. Cork', *Journal of the Cork Historical and Archaeological Society*, Second Series XIX: 97, 1-8.

Burns, G. (1995), *If Only: Historical Sketches of the Belcoo Area* (Florencecourt: No publisher).

Byrne, D. R. (2005), *Vintage Port. A View of Portrush using Hitherto Largely Unseen Photographic and other Material* (Portrush: Ballywillan Drama Group).

Cant, S. (2005), 'Understanding why people use complementary and alternative medicine', in T. Heller, G. Lee-Treweek, J. Katz, J. Stone and S. Spurr (eds) *Perspectives on Complementary and Alternative Medicine* (Abingdon: Routledge, Taylor & Francis and The Open University) pp. 173-204.

Cantor, D. (1990), 'The contradictions of specialization: Rheumatism and the decline of the Spa in inter-war Britain', in R. Porter (ed.) *Medical History, Supplement No. 10.* (London: Wellcome Institute for the History of Medicine) pp. 127-144.

Carleton, W. (1860), *The Evil Eye or The Black Spectre* (Dublin: James Duffy).

Carroll, J. and Tuohy, P. (1999), *Village by Shannon. The Story of Castleconnell and its Hinterland* (Limerick: No publisher).

Carroll, M. P. (1999), *Irish Pilgrimage: Holy Wells and Popular Catholic Devotion* (Baltimore: The Johns Hopkins University Press).

Clancy, M., Clancy, E., Forde, P. and Clancy, P. (2003), *Ballinaglera and Inismagrath: The History and Tradition of Two Leitrim Parishes* (Ballinaglera: No publisher).

Clare County Library (2009), '*Clare Places: Towns and Villages. Lahinch*' www. clarelibrary.ie/eolas/coclare/places/lahinch.htm. Accessed 4 October.

Clare, L. (1998), *Victorian Bray: A Town Adapts to Changing Times* (Dublin: Irish Academic Press).

Clarke, D., Doel, M. and Segrott, J. (2004), 'No alternative? The regulation and professionalization of complementary and alternative medicine in the United Kingdom', *Health & Place* 10: 329-338.

Cliff House Hotel (2009), *Cliff House Hotel Brochure* (Ardmore: Cliff House Hotel).

Cloke, P. and Jones, O. (2001), 'Dwelling, place, and landscape: An orchard in Somerset', *Environment and Planning A* 33:4, 649-666.

Cobbett, W. (1967) [1830], *Rural Rides* (Harmondsworth: Penguin).

Coccheri, S., Gasberrini, G., Valenti, M., Nappi, G. and Di Orio, F. (2008), 'Has time come for a re-assessment of spa therapy? The NAIADE survey in Italy', *International Journal of Biometeorology* 52: 231-237.

Coley, N., G. (1990), 'Physicians, Chemists and the Analysis of Mineral Waters: "The most difficult part of Chemistry"', in R. Porter (ed.) *Medical History, Supplement No. 10.* (London: Wellcome Institute for the History of Medicine) pp. 56-66.

Collins, D. and Kearns, R. (2007), 'Ambiguous landscapes: Sun, risk and recreation on New Zealand beaches', in A. Williams (ed.) *Therapeutic Landscapes* (Aldershot: Ashgate) pp. 15-32.

Condren, M. (2002), *The Serpent and the Goddess* (Dublin: New Island Books).

Conlon, L. (1999) 'The Holy Wells of County Louth', *Journal of the County Louth Archaeological and Historical Society* 24:3 pp.329-345.

Connell, J. (2006), 'Medical tourism: Sea, sun, sand and ... surgery', *Tourism Management* 27: 1093-1100.

Conradson, D. (2005a), 'Freedom, space and perspective: Moving encounters with other ecologies', in J. Davidson, L. Bondi and M. Smith (eds) *Emotional Geographies* (Aldershot: Ashgate) pp. 103-116.

Conradson, D. (2005b), 'Landscape, care and the relational self: Therapeutic encounters in rural England', *Health & Place* 11: 337-348.

Conradson, D. (2007), 'The experiential economy of stillness: Places of retreat in contemporary Britain', in A. Williams (ed.) *Therapeutic Landscapes* (Aldershot: Ashgate) pp. 33-48.

Corbin, A. (1994), *The Lure of the Sea. The Discovery of the Seaside in the Western World 1750-1840* (Cambridge: Polity).

Cossick, A. and Galliou, P. (eds) (2006), *Spas in Britain and in France in the Eighteenth and Nineteenth Centuries* (Newcastle-upon-Tyne: Cambridge Scholars).

Croker, T. C. (1824), *Researches in the South of Ireland: Illustrative of the Scenery, Architectural Remains, and the Manners and Superstitions of the Peasantry* (London: John Murray).

Cronin, S. (2008), 'Have spas run out of steam?', *Irish Independent*, 2 September.

Crouch, D. (2000), 'Places around us: Embodied lay geographies in leisure and tourism', *Leisure Studies* 19: 63-76.

Croutier, A. L. (1992), *Taking the Waters. Spirit, Art, Sensuality* (New York: Abbeville Press).

Cummins, S., Curtis, S., Diez-Roux, A. and Macintyre, S. (2007), 'Understanding and representing "place" in health research: A relational approach', *Social Science and Medicine* 65: 1825-1838.

Curtis, L. P. (1997), *Apes and Angels. The Irishman in Victorian Caricature* (Washington: Smithsonian Institution Press).

Curtis, S. (2004), *Health and Inequalities* (London: Sage).

Curtis, S., Gesler, W. M., Priebe, S. and Francis, S. (2009), 'New spaces of inpatient care for people with mental illness: A complex "rebirth" of the clinic? ', *Health & Place* 15: 340-348.

Davidson, J. and Milligan, C. (2004), 'Embodying emotion sensing space: Introducing emotional geographies', *Social & Cultural Geography* 5:4, 523-532.

Davidson, J. and Parr, H. (2007), 'Anxious subjectivities and spaces of care: Therapeutic geographies of the UK National Phobic Society', in A. Williams (ed.) *Therapeutic Landscapes* (Aldershot: Ashgate) pp. 95-110.

Davies, K. M. (1993), 'For health and pleasure in the British fashion: Bray, Co. Wicklow, as a tourist resort, 1750-1914', in M. Cronin and R. O'Connor (eds) *Tourism and Ireland: A Critical Analysis* (Cork: Cork University Press) pp. 29-45.

Davies, K. M. (2007), *That Favourite Resort: The story of Bray, Co. Wicklow* (Bray: Wordwell).

de Bovet, M. A. (1891), *Three Months' Tour in Ireland* (London: Chapman and Hall).

De Latocnaye, C. (1984), *A Frenchman's Walk through Ireland 1796–7* (Belfast: Blackstaff Press).

DeMiglio, L. and Williams, A. (2008), 'A Sense of Place, A Sense of Well-being', in J. Eyles and A. Williams (eds) *Sense of Place, Health and Quality of Life* (Aldershot: Ashgate) pp. 15-30.

Denbeigh, K. (1981), *A Hundred British Spas* (London: Spa Publications).

Department of the Environment, Heritage and Local Government (2008), *Sites and Monuments Record for the Republic of Ireland* (Dublin: DOEHLG).

De Vál, S. (ed.) (2007), *Our Lady's Island, Oileán Mhuire, Insula Beatae Mariae* (Our Lady's Island: Dennis Brennan).

de Verteuil, G. and Andrews, G. (2007), 'Surviving profoundly unhealthy places: The ambivalent, fragile and absent therapeutic landscapes of the Soviet Gulag', in A. Williams (ed.) *Therapeutic Landscapes* (Aldershot: Ashgate) pp. 273-287.

Doel, M. and Segrott, J. (2003), 'Beyond belief? Consumer culture, complementary medicine, and the dis-ease of everyday life', *Environment and Planning D* 21: 739-759.

Dooley-Shannon, M. (1998), '*The Historical Development of Tourism in Clare: Portrait of a Spa Town - Lisdoonvarna, 1800-1914*', MA Thesis, University of Limerick, Limerick.

Doran, A. L. (1903), *Bray and Environs* (Bray: No publisher).

Dorn, M. and Laws, G. (1994), 'Social theory, body politics, and medical geography: Extending Kearns's invitation', *The Professional Geographer* 46:1, 106-110.

Duffy, P. (2007), *Exploring the History and Heritage of Irish Landscapes* (Dublin: Four Courts Press).

Dunkley, C. M. (2009), 'A therapeutic taskscape: Theorizing place-making, discipline and care at a camp for troubled youth', *Health & Place* 15: 88-96.

Durie, A. (2003a), 'Medicine, health and economic development: Promoting spa and seaside resorts in Scotland c. 1750-1830', *Medical History* 47: 195-216.

Durie, A. (2003b), *Scotland for the Holidays: Tourism in Scotland c1780-1939* (East Linton: Tuckwell).

Durie, A. (2006), *Water is Best: The Hydros and Health Tourism in Scotland 1840-1940* (Edinburgh: John Donald).

Dwyer, P. (1998) [1876], *A Handbook to Lisdoonvarna and its Vicinity* (Ennis: Clasp Press).

Egan, P.M. (1894) *History, Guide & Directory of County and City of Waterford* (Kilkenny: P.M. Egan).

Einwalter, D. (2007), 'Reclaiming the therapeutic value of public space through roadside art and memorials in rural Nevada', in A. Williams (ed.) *Therapeutic Landscapes* (Aldershot: Ashgate) pp. 333-348.

Eliade, M. (1961), *Images and Symbols: Studies in Religious Symbolism* (Mission: Sheed, Andrews and McMeel).

English, J., Wilson, K. and Keller-Olaman, S. (2008), 'Health, healing and recovery: Therapeutic landscapes and the everyday lives of breast cancer survivors', *Social Science and Medicine* 67: 68-78.

Evans, E. E. (1957), *Irish Folk Ways* (London: Routledge & Kegan Paul).

Eyles, J. (1985), *Senses of Place* (Warrington: Silverbrook Press).

Eyles, J. and Williams, A. (eds) (2008), *Sense of Place, Health and Quality of Life* (Aldershot: Ashgate).

Fáilte Ireland (2007), *Health and Wellness. Positioning Strategy for Key Markets* (Dublin: Fáilte Ireland).

Farrant, S. (1987), 'London by the sea: Resort development on the south coast of England, 1880-1939', *Journal of Contemporary History* 22, 137-162.

Ferrar, J. (1767), *An History of the City of Limerick* (Limerick: Andrew Welsh).

Finlan, M. (1993), 'Seaweed Baths Enniscrone', *Ireland of the Welcomes*, 42: 36-38.

Fitzgerald, W. (1914), 'Father Moore's Well', *Journal of the County Kildare Archaeological Society* 7:5, 329-332.

Fleetwood, J. F. (1993), 'Dublin private medical schools in the nineteenth century', *Dublin Historical Record* 46:1, 31-45.

Flinn, D.E. (1888) *Ireland. Its Health Resorts and Watering Places* (London: Kegan Paul Trench and Co).

Flynn, A. (2004) *Bray in Old Photographs* (Dublin: Gill and Macmillan).

Foster, R. F. (1988), *Modern Ireland 1600-1972* (Harmonsworth: Penguin).

Foucault, M. (1976) [1963], *The Birth of the Clinic: An Archaeology of Medical Perception* (London: Tavistock).

Foucault, M. (1977) [1975], *Discipline and Punish: The Birth of the Prison* (Paris: Gallimard).

Frost, G. (2004), 'The spa as a model of an optimal healing environment', *The Journal of Complementary and Alternative Medicine* 10: Supplement 1, S85-S92.

Furlong, I. (2006), 'The saga of Lisdoonvarna – "From Queen of Irish Spas" to modern matchmaking mecca', in A. Cossick and P. Galliou (eds) *Spas in Britain and in France in the eighteenth and nineteenth centuries* (Newcastle-upon-Tyne: Cambridge Scholars) pp. 239-261.

Furlong, I. (2009), *Irish Tourism, 1880-1980* (Dublin: Irish Academic Press).

Gattrell, A. and Elliott, S. (2009), *Geographies of Health: An Introduction* (Chichester: Wiley-Blackwell).

Gell, A. (1988), *Art and Agency: An Anthropological Theory* (Oxford: Clarendon Press).

Geores, M. E. (1998), 'Surviving on metaphor: How 'Health = Hot Springs' created and sustained a town', in R. Kearns and W. M. Gesler (eds) *Putting*

Health into Place: Landscape, Identity and Well-Being (Syracuse: Syracuse University Press) pp. 36-52.

Gerten, D. (2008), 'Water of life, water of death: Pagan notions of water from antiquity', in S. Shaw and A. Francis (eds) *Deep Blue: Critical Reflections on Nature, Religion and Water* (London: Equinox) pp. 33-48.

Gesler, W. M. (1992), 'Therapeutic landscapes: Medical Issues in the light of the new cultural geography', *Social Science and Medicine* 34:7, 735-746.

Gesler, W. M. (1993), 'Therapeutic landscapes: Theory and a case study of Epidauros, Greece', *Environment and Planning D* 11:2, 171-189.

Gesler, W. M. (1996), 'Lourdes: Healing in a place of pilgrimage', *Health & Place* 2:2, 95-105.

Gesler, W. M. (1998), 'Bath's reputation as healing place', in R. Kearns and W. Gesler. (eds) *Putting Health into Place: Landscape, Identity and Well-Being* (Syracuse: Syracuse University Press) pp. 17-35.

Gesler, W. M. (2003), *Healing Places* (Lanham: Rowman & Littlefield).

Gesler, W. M. and Kearns, R. (2002), *Culture/Place/Health* (London: Routledge).

Gibson, C. and Connell, J. (2005), *Music and Tourism: On the Road Again* (Clevedon: Channel View Publications).

Gibson, C. B. (1861), *History of the County and City of Cork. Volumes 1 and 2* (London: Thomas C. Newby).

Graber, L. H. (1976), *Wilderness as Sacred Space* (Washington: Association of American Geographers).

Granville, A. B. (1971) [1841], *Spas of England and Principal Sea-bathing Places* (Bath: Adams and Dart).

Graves, R. (1948), *The White Goddess: A Historical Grammer of Poetic Myth* (New York: The Noonday Press).

Guerra, F. (1966), 'Aztec Medicine', *Medical History* 10:4, 315-338.

Guiry, M. (2006) 'Seaweed in Ireland', in A. T. Critchley, Ohno, M. and D. B. Largo (eds) *World Seaweed Resources. An Authoritative Reference System.* (Wokingham: ETI Information Services).

Hackett, M. (1994), *Youghal's Fading Footsteps: The Humour, History & People of East Cork and West Waterford* (Blarney: On Stream).

Haggerty, B. (2007), *The Holy Wells of Ireland* http://www.irishcultureandcustoms. com/ALandmks/HolyWells.html. Accessed 17 July.

Hahn, G. and Schoenfels, H.-K. (1986), *Von der heilkraft des wassers: Eine kulturgeschichte der Brunnen und Bäder* (Augsburg: Weltbild-Bücherdienst).

Hall, S. C. and Hall, S. C (1843), *Ireland, its Scenery and Character ...* (London: How & Parsons).

Hamlin, C. (1990), 'Chemistry, medicine and the legitimization of English spas, 1740-1840', in R. Porter (ed.) *Medical History, Supplement No. 10.* (London: Wellcome Institute for the History of Medicine) pp. 67-81.

Handcock, W. D. (1899), *The History and Antiquities of Tallaght* (Dublin, No publisher).

Hannon, K. (1983), 'Castleconnell: Part one', *Old Limerick Journal* 15: 23-28.

Hannon, K. (1984), 'Castleconnell: Part two', *Old Limerick Journal* 16: 5-9.

Harbison, P. (1991), *Pilgrimage in Ireland: The Monuments and the People* (London: Barrie and Jenkins).

Hardy, P. D. (1836), *The Holy Wells of Ireland* (Dublin, Hardy and Walker).

Harley, D. (1990), 'A Sword in a Madman's Hand: Professional opposition to popular consumption in the waters literature of Southern England and the Midlands, 1570-1870', in R. Porter (ed.) *Medical History, Supplement No. 10.* (London: Wellcome Institute for the History of Medicine) pp. 48-55.

Harris, W. and Smith, C. (1744), *The Antient and Present state of the County of Down* (Dublin: A. Reilly).

Harvey, D. (1996), *Justice, Nature, and the Geography of Difference* (Oxford: Blackwell).

Hassan, J. (2003), *The Seaside, Health and the Environment in England and Wales since 1800* (Aldershot: Ashgate).

Healy, E. (2001), *In Search of Ireland's Holy Wells* (Dublin: Wolfhound Press).

Heller, T. (2005), 'Complementary and alternative medicine and mental health', in T. Heller, G. Lee-Treweek, J. Katz, J. Stone and S. Spurr (eds) *Perspectives on Complementary and Alternative Medicine* (Abingdon: Routledge, Taylor & Francis and The Open University) pp. 111-139.

Heller, T., Lee-Treweek, G., Katz, J., Stone, J. and Spurr, S. (eds) (2005), *Perspectives on Complementary and Alternative Medicine* (Abingdon: Routledge, Taylor & Francis and the Open University).

Hembry, P. (1990), *The English Spa 1560-1815. A Social History* (London: The Athlone Press).

Hembry, P. (1997), *British Spas From 1815 to the Present. A Social History* (London: The Athlone Press).

Henchy, P. (1958), 'A Bibliography of Irish Spas', *The Bibliographical Society of Ireland*, 6:7, 98-111.

Henry, W. (1739), *Hints towards a Natural and Topographical History of the Counties of Sligoe, Donegal, Fermanagh and Lough Erne* (MIC 198/1) (Belfast: Public Records Office of Northern Ireland) pp.10-11.

Herbert, R. (1948), 'Castleconnell and its Spa', *North Munster Antiquarian Journal* 5: 117-140.

Hetherington, K. and Law, J. (2000), 'Guest editorial', *Environment and Planning D* 18: 127-132.

Heuston, J. (1993), 'Kilkee – The origins and development of a west coast resort.', in M. Cronin and R. O'Connor (ed.) *Tourism and Ireland: A Critical Analysis* (Cork: Cork University Press) pp. 13-28.

Hoey, B. A. (2007), 'Therapeutic uses of place in the intentional space of purposive community', in A. Williams (ed.) *Therapeutic Landscapes* (Aldershot: Ashgate) pp. 297-314.

Hogan, H. (1842), *A Directory of Kilkee* (Limerick: K.M. Goggin).

Holloway, J. (2006), 'Enchanted spaces: The sèance, affect and geographies of religion', *Annals of the Association of American Geographers* 96:1, 182-187.

Horgan, D. (2002), *The Victorian Visitor in Ireland: Irish Tourism 1840-1980* (Cork: Heritage Council).

Hoyez, A.-C. (2007a), 'The "world of yoga": The production and reproduction of therapeutic landscapes', *Social Science and Medicine* 65: 112-124.

Hoyez, A.-C. (2007b), 'From Rishikesh to Yogaville: The globalization of therapeutic landscapes', in A. Williams (ed.) *Therapeutic Landscapes* (Aldershot: Ashgate) pp. 49-64.

Hubbard, P. (2005), 'Space/Place', in D. Atkinson, P. Jackson, D. Sibley and N. Washbourne (eds) *Cultural Geography. A Critical Dictionary of Key Concepts* (London: I.B. Tauris) pp. 41-48.

Illich, I. (1986), *H2O and the Waters of Forgetfulness* (London: Marion Boyars).

Inchydoney Island Lodge and Spa (2008), *Inchydoney Island Lodge and Spa Brochure* (Inchydoney: Inchydoney Island Lodge and Spa).

Inglis, H. D. (1835), *Ireland in 1834: A Journey Throughout Ireland, during the Spring, Summer, and Autumn of 1834* (London: Whittaker).

Ingold, T. (2000), *The Perception of the Environment: Essays in Livelihood, Dwelling and Skill* (London: Routledge).

Irish Builder (1861), 'Our watering places. Portrush-Bangor-Holywood', *Irish Builder* Vol. IX:182, 175.

Jackson, P. (1989), *Maps of Meaning: An Introduction to Cultural Geography* (London: Unwin Hyman).

Jackson, R. (1990), 'Waters and Spas in the Classical World', in R. Porter (ed.) *Medical History, Supplement No. 10.* (London: Wellcome Institute for the History of Medicine) pp. 1-13.

Jephson, M. D. (1964), *An Anglo-Irish Miscellany. Some Records of the Jephsons of Mallow* (Dublin Allen Figgis).

Jones, Bryony, (2006), *The City that got its Soul back.* http://news.bbc.co.uk/2/hi/uk_news/england/somerset/5225872.stm. Accessed 31 July.

Jones, G. and Malcolm, E. (eds) (1999), *Medicine, Disease and the State in Ireland, 1650-1940* (Cork: Cork University Press).

Jones, O. and Cloke, P. (2002), *Tree Cultures: The Place of Trees and Trees in their Place* (Oxford, Berg).

Joseph, A., Kearns, R. and Moon, G. (2009), 'Recycling former psychiatric hospitals in New Zealand: Echoes of deinstitutionalisation and restructuring', *Health & Place* 15: 79-87.

Joseph, A. and Phillips, D. (1984), *Accessibility and Utilization: Geographical Perspectives on Health Care Delivery* (New York: Harper and Row).

Joyce, J. (1997) [1922], *Ulysses* (London: Picador).

Joyce, W. S. J. (1901), *Lucan and its Neighbourhood* (Dublin: M.H. Gill and Son).

Kavita, P. (2002), 'Race, Class and the Imperial Politics of Ethnography in India, Ireland and London, 1850–1910', *Irish Studies Review* 10:3, 289-302.

Kearns, G. (2007), 'The History of Medical Geography after Foucault', in Crampton, J. W. and S. Elden (eds) *Space, Knowledge and Power: Foucault and Geography* (Aldershot: Ashgate) pp. 205-222.

Kearns, R. and Barnett, J. R. (1999), 'Auckland's Starship Enterprise: Placing Metaphor in a Children's Hospital', in A. Williams (ed.) *Therapeutic Landscapes: The Dynamic between Place and Wellness* (Lanham: University Press of America) pp. 169-199.

Kearns, R. and Gesler, W. M. (eds) (1998), *Putting Health into Place: Landscape, Identity and Well-Being* (Syracuse: Syracuse University Press).

Kearns, R. and Moon, G. (2002), 'From medical to health geography: novelty, place and theory after a decade of change', *Progress in Human Geography* 26:5, 605-625.

Keena, M. (2007), *The Road to Lisdoonvarna* (Lisdoonvarna: Wolf Hill Publishing).

Kelly, B., Kelly, V. and Foster, R. (1995), *The Book of Kelly's* (Dublin: Zeus Publishing).

Kelly, J. (2009), '"Drinking the Waters": Balneotherapeutic medicine in Ireland, 1660-1850. Medicine in 17th and 18th century Ireland', *Studia Hibernica*, 35: 99-145.

Kelly's Resort Hotel (2009), *SeaSpa Brochure* (Rosslare: Kelly's Resort Hotel).

Kingsley Hotel (2009), *Yauvana Spa Brochure* (Cork: Kingsley Hotel).

Kingsley, J. Y., Townsend, M., Phillips, R. and Aldous, D. (2009), '"If the land is healthy ... it makes the people healthy": The relationship between caring for Country and health for the YortaYorta Nation, Boonwurrung and Bangerang Tribes', *Health & Place* 15: 291-299.

Kleinman, A. (1973), 'Symbolic reality: On a central problem in the philosophy of medical inquiry', *Inquiry* 16: 206-213.

Knott, K. and Franks, M. (2007), 'Secular values and the location of religion: A spatial analysis of an English medical centre', *Health & Place* 13: 224-237.

Knott, M. J. (1997) [1836], *Three Months at Kilkee* (Ennis: CLASP Press).

Knox, A. (1845), *The Irish Watering Places: Their Climate, Scenery and Accommodations* (Dublin: William Curry).

Lea, J. (2006), 'Experiencing festival bodies: Connecting massage and wellness', *Tourism Recreation Research* 31(1): 57-66.

Lea, J. (2008), 'Retreating to nature: Rethinking "therapeutic landscapes"', *Area* 40(1): 90-98.

Lee, G. (2004), *Spa Style Europe. Therapies, Cuisines, Spas* (Singapore: Editions Didier Millet/Weatherhill).

Lenček, L. and Bosker, G. (1998), *The Beach. The History of Paradise on Earth* (London, Secker & Warburg).

Lewis, S. (1837), *Lewis' Topographical Dictionary of Ireland. Volumes I and II* (London: S. Lewis and Co.)

Lincoln, S. (2000), *Ardmore: Memory and Story* (Ardmore: No publisher).

Logan, P. (1972), *Making the Cure: A Look at Irish Folk Medicine* (Dublin: The Talbot Press).

Logan, P. (1981), *Irish Country Cures* (Belfast: Appletree Press).

Lorimer, H. (2003), 'Telling small stories: Spaces of knowledge and the practice of geography', *Transactions of the Institute of British Geographers* 28: 197-217.

Lorimer, H. (2005), 'Cultural geography: The busyness of being "more-than-representational"', *Progress in Human Geography* 29:1, 83-94.

Lorimer, H. (2008), 'Cultural geography: Non-representational conditions and concerns', *Progress in Human Geography* 32:4, 551-559.

Lynd, R. (1998) [1911], 'Rambles in Clare, 1911', in Ó Dálaigh, B. (ed.) *The Strangers Gaze: Travels in County Clare, 1534-1950* (Ennis: CLASP Press) pp. 325-329.

Mackaman, D. P. (1998), *Leisure Settings: Bourgeois Culture, Medicine and the Spa in Modern France* (Chicago: University of Chicago Press).

MacNeill, M. (1982), *The Festival at Lughnasa* (Dublin: UCD Folklore Department).

MacNeill, M. (1988), 'Ritual horse-bathing at harvest time', *Béaloideas* 56: 93-96.

Madden, T. M. (1891), *Lucan Spa and Hydropathic as a Modern Health Resort* (Dublin: M.H. Gill & Son).

Magennis, E. (2002), '"A land of milk and honey": The Physico-Historical Society, improvement and the surveys of mid-eighteenth century Ireland', *Proceedings of the Royal Irish Academy* 102:C, 199-217.

Marrinan, S. (1982), 'Kilkee as it was a hundred years ago', *The Other Clare* 6: 41-42.

Matless, D. (2000), 'Action and noise over a hundred years: The making of a nature region', *Body and Society* 6: 141-165.

McClintock, A. (1995), *Imperial Leather: Race, Gender and Sexuality in the Colonial Context* (London: Routledge).

McCormack, D. (2004), 'An event of geographical ethics in spaces of affect', *Transactions of the Institute of British Geographers* 28: 488-507.

McCullough, S. (1968), *Ballynahinch, Centre of Down* (Ballynahinch: Ballynahinch Chamber of Commerce).

McKeown. T. (1979), *The Role of Medicine: Dream, Mirage or Nemesis?* (Oxford: Blackwell).

McKinlay, J. M. (1893), *Folklore of Scottish Lochs and Springs* (Glasgow: W. Hodge & Co).

Meinig, Donald (1979), *The Interpretation of Ordinary Landscapes: Geographical Essays* (New York: Oxford University Press).

Merleau-Ponty, M. (1962) [1942], *Phenomenology of Perception* (London: Routledge and Kegan Paul).

Milligan, C. (2007), 'Restoration or risk? Exploring the place of the common place', in A. Williams (ed.) *Therapeutic Landscapes* (Aldershot: Ashgate) pp. 255-272.

Milligan, C. and Bingley, A. (2007), 'Restorative places or scary spaces? The impact of woodland on the mental well-being of young adults', *Health & Place* 13: 799-811.

Milligan, S. F. (1889), 'The ancient Irish hot-air bath', *Journal of the Royal Society of Antiquaries of Ireland* XIX: 4, 268-270.

Mitchell, D. (2000), *Cultural Geography. A Critical Introduction* (Oxford: Blackwell).

Monart Spa (2008), *Monart Spa Brochure* (Enniscorthy: Monart Spa).

Mulcahy, D. B. (1891), 'An ancient Irish hot-Air bath, or sweat-house, on the island of Rathlin', *Journal of the Royal Society of Antiquaries of Ireland* XXI, 7: 589-590.

Murphy, D.J. (1979), Halcyon days of St. Ann's Hill Hydro, *Cork Evening Echo*, 17 December.

Murphy, Fr. I. (1977), 'At the seaside in Kilkee in the 1830s and 1840s', *The Other Clare* 1: 26-31.

Murphy, P. (2008), *The Business of Resort Management* (London: Butterworth-Heinmann).

Myers, K. (1984), 'The Mallow Spa', *Mallow Field Club Journal* 2: 5-19.

Nash, C. (2000), 'Performativity in practice: Some recent work in cultural geography', *Progress in Human Geography* 24:4, 653-664.

National Folklore Collection (1934a), Manuscript 466, *Cúntaisí ar Thoibreacha Beannuithe do fuairtear ó Mhúinteorí ar fuaid Chúige Mumhan* (Dublin: National Folklore Collection).

National Folklore Collection (1934b), Manuscript 467, *Cúntaisí ar Thoibreacha Beannuithe do fuairtear ó Mhúinteorí ar fuaid Chúige Chonnacht agus Uladh* (Dublin: National Folklore Collection).

National Folklore Collection (1934c), Manuscript 468, *Cúntaisí ar Thoibreacha Beannuithe do fuairtear ó Mhúinteorí ar fuaid Chúige Laighean* (Dublin: National Folklore Collection).

National Folklore Collection (1938), *National Folklore Collection Schools Manuscript 703*: 159; Eileen Coulds (14), Clonmellon, County Meath. 2 June, 1938.

Nolan, M. L. and Nolan, S. (1992), 'Religious sites as tourism attractions in Europe', *Annals of Tourism Research* 19: 68-78.

Northern Ireland Environmental Agency (2009), *Sites and Monuments Record for Northern Ireland* (Belfast: NIEA).

Ó'Cadhla, S. (2002), *The Holy Well Tradition. The Pattern of St Declan, Ardmore, County Waterford, 1800-2000* (Maynooth: Four Courts Press).

O'Carroll, C. (1987), 'Tourist Development in 19th Century County Clare', *The Other Clare* 11: 26-28.

O'Connell, P. (1957), 'Castle Kieran', *Ríocht na Midhe* 1(3): 17-33.

O'Connell, P. (1963), 'A Co. Cavan Itinerary circa 1744', *Breifne: Journal of the Brefine Historical Society* 2(6): 254-273.

O'Connor, K. (1999), *Ironing the Land. The Coming of the Railways to Ireland* (Dublin: Gill and Macmillan).

O'Donohue, John (1997), *Anam Chara. Spiritual Wisdom from the Celtic World* (London: Bantam).

Ò'Duinn, S. (2005), *The Rites of Brigid: Goddess and Saint* (Dublin: The Columba Press).

Ó'Fearghaill, F. (1987), 'From Tuam to Ballyspellan', *Old Kilkenny Review: Journal of the Kilkenny Archaeological Society* 3: 399-405.

O'Leary, S. (2000), 'St. Ann's Hydro', *Old Blarney: Journal of the Blarney and District Historical Society* 5: 3-31.

O'Mahoney, C. (2006), 'Bubble and chic', *Sunday Tribune*, 19 March.

O'Mahoney, C. (1986), *The Maritime Gateway to Cork. A history of the outports of Passage West and Monkstown from 1754-1942* (Cork: Tower Books).

O'Reilly, S. (1988), 'Forgotten holy wells', *Mallow Field Club Journal* 5: 128-132.

Orme, A. R. (1966), 'Youghal, County Cork', *Irish Geography* 5: 121-149.

O'Sullivan, S. (1977), 'Jonathan Swift and Wexford's spa', *Journal of Old Wexford Society* 6: 63-68.

Palka, E. (1999), 'Accessible wilderness as a therapeutic landscape: Experiencing the nature of Denali National Park', in A. Williams (ed.) *Therapeutic Landscapes: The Dynamic between Place and Wellness*. (Lanham: University Press of America) pp. 29-51.

Palmer, R. (1990), '"In this our lightye and learned tyme": Italian Baths in the era of the Renaissance', in R. Porter (ed.) *Medical History, Supplement No. 10*. (London: Wellcome Institute for the History of Medicine) pp. 14-22.

Parr, H. (2002), 'New body-geographies: The embodied spaces of health and medical information on the Internet', *Environment and Planning A* 20: 73-95.

Paterson, M. (2005), 'Affecting touch: Towards a "Felt" Phenomenology of Human Touch', in J. Davidson, L. Bondi and M. Smith (eds) *Emotional Geographies* (Aldershot: Ashgate) pp. 161-173.

Pearsell, J. (ed.) (2001), *The New Oxford Dictionary of English* (Oxford: Oxford University Press).

Petroune, I. and Yachina, E. (2009), 'Heritage of spa and health tourism in Russia', in M. Smith and L. Puzcko (eds) *Health and Wellness Tourism* (Oxford: Butterworth-Heinemann) pp. 285-290.

Philo, C. (2000), 'Foucault's geography', in M. Crang and N. Thrift (eds) *Thinking Space* (London: Routledge) pp. 205-238.

Philo, C. and Parr, H. (2003), 'Introducing psychoanalytic geographies', *Social & Cultural Geography* 4:3, 283-293.

Pinfold, V. (2000), 'Building up safe havens ... all around the world': Users' experiences of living in the community with mental health problems', *Health & Place* 6:3, 201-212.

Pococke, R. (1995) [1752], *Richard Pococke's Irish Tours* (Dublin: Irish Academic Press).

Porter, R. (1989), *Health for sale: Quackery in England 1660-1850* (Manchester: Manchester University Press).

Porter, R. (1990), 'The medical history of waters and spas: Introduction', in R. Porter (ed.) *Medical History, Supplement No. 10*. (London: Wellcome Institute for the History of Medicine) pp. vii-xii.

Porter, R. (1992), 'The patient in England, c.1660–c.1800', in A. Wear. (ed.) *Medicine in Society. Historical Essays* (Cambridge: Cambridge University Press) pp. 91-118.

Porter, R. (1999), *The Greatest Benefit to Mankind: A Medical History of Humanity from Antiquity to the Present* (London: Fontana).

Price, R. (1981), 'Hydropathy in England, 1840-1870', *Medical History* 25: 269-280.

Pringle, D. and Houghton, F. (2007), 'Health and disease in Ireland', in B. Bartley and R. Kitchin. (eds) *Understanding Contemporary Ireland* (London: Pluto Press) pp. 279-288.

Quinn, S. (2003), *Sea Baths of South County Dublin* (Dublin: Foxrock Local History Club Publication No. 51).

Rattue, J. (1995), *The Living Stream: Holy Wells in Historical Context* (Woodbridge: The Boydell Press).

Ràtz, T. (2009), 'Hot springs in Japanese domestic and international tourism', in M. Smith and L. Puzcko (eds) *Health and Wellness Tourism* (Oxford: Butterworth-Heinemann) pp. 345-349.

Relph, E. (1976), *Place and Placelessness* (London: Pion).

Richardson, P. (1939), 'Sweathouses between Blacklion and Dowra, County Cavan', *Ulster Journal of Archaeology* 2: 3rd Series, 30-35.

Rickard, A. and O'Callaghan, L. (2001), *FishStoneWater. Holy Wells of Ireland* (Cork: Atrium).

Rockel, I. (1986), *Taking the Waters. Early Spas in New Zealand* (Wellington, Government Printing Office).

Rojas Alba, H. (1996), Temazcal. *Tlahui-Medic* 2 (II), http://www.tlahui.com/temaz1.html. Accessed 28 January.

Rose, G. (1993), *Feminism and Geography: The Limits of Geographical Knowledge* (Cambridge: Polity Press).

Rose, M. and Wylie, J. (2006), 'Animating landscape', *Environment and Planning D* 24: 475-479.

Rutty, J. (1757), *An Essay Towards a Natural, Experimental and Medicinal History of the Mineral Waters of Ireland* (Dublin: No publisher).

Ryan, M., D. (1824), *Treatise on the Most Celebrated Mineral Waters of Ireland* (Kilkenny: J. Reynolds).

Said, E. (1985), *Orientalism* (Harmondsworth: Penguin).

Saks, M. (2005), 'Political and Historical Perspectives', in T. Heller, G. Lee-Treweek, J. Katz, J. Stone and S. Spurr (eds) *Perspectives on Complementary and Alternative Medicine* (Abingdon: Routledge, Taylor & Francis and The Open University) pp. 59-82.

Sellner, E. C. (2004), *Pilgrimage* (South Bend: Sorin Books).

Shackley, M. (2001), *Managing Sacred Sites* (London: Continuum).

Sharpe, R. (2006), 'Five years late, £30m overspent, mired in legal rows. Finally, Bath Spa opens', *The Observer*, 30 July.

Shaw, S. and Francis, A. (eds) (2008), *Deep Blue: Critical Reflections on Nature, Religion and Water* (London: Equinox).

Sheehan, J. (1995), St. Ann's Hydro – Cork's Turkish Baths. *Cork Evening Echo*, 29 September.

Shields, R. (1991), *Places on the Margin: Alternative Geographies of Modernity* (London: Routledge).

Shifrin, M. (2009), '*Victorian Turkish baths: Their origin, development, & gradual decline.*' http://www.victorianturkishbath.org/. Accessed 12 February.

Sibley, D. (2003), 'Geography and psychoanalysis: Tensions and possibilities', *Social & Cultural Geography* 4:3, 391-399.

Sidaway, J. (2000), 'Postcolonial geographies: An exploratory essay', *Progress in Human Geography* 24:4, 591-612.

Simon, B. (2000), 'Tree Traditions and Folklore from Northeast Ireland', *Arboricultural Journal* 24: 15-40.

Singh, R. P. B. (2006), 'Pilgrimage in Hinduism: Historical context and modern perspectives', in D. J. Timothy and D. H. Olsen (eds) *Tourism, Religion and Spiritual Journey* (Abingdon: Routledge) pp. 220-236.

Slater, E. (2007), 'Reconstructing "Nature" as a picturesque theme park: the colonial case of Ireland', *Early Popular Visual Culture* 5:3, 231-245.

Smith, M. (2003), *Issues in Cultural Tourism* (Abingdon: Routledge).

Smith, M. (2005), 'On "Being" Moved by Nature: Geography, Emotion and Environmental Ethics', in J. Davidson, L. Bondi and M. Smith (eds) *Emotional Geographies* (Aldershot: Ashgate) pp. 219-230.

Smith, M. and Kelly, C. (2006), 'Wellness Tourism', *Tourism Recreation Research* 31:1, 1-4.

Smith, M. and Puczko, L. (eds) (2009), *Health and Wellness Tourism* (Oxford: Butterworth-Heinemann).

Smollett, T. (1984) [1771], *The Expedition of Humphrey Clinker* (Oxford: Oxford University Press).

Smyth, F. (2005), 'Medical geography: Therapeutic places, spaces and networks', *Progress in Human Geography* 29:4, 488-495.

Spa-Ireland.Com (2009), http://www.spa-ireland.com/. Accessed 25 April.

Sperrin Tourism (2008), *Discover the Sperrins and Make it Yours* (Moneynore: Sperrins Tourism.).

Stallybrass, P. and White, A. (1986), *The Politics and Poetics of Transgression* (London: Methuen).

St. Leger, A. (1994), *Youghal: Historic Walled Port. The Story of Youghal* (Youghal: Youghal Urban District Council).

Stone, J. and Katz. J. (2005), 'The therapeutic relationship and complementary and alternative medicine', in T. Heller, G. Lee-Treweek, J. Katz, J. Stone and S. Spurr (eds) *Perspectives on Complementary and Alternative Medicine* (Abingdon: Routledge, Taylor & Francis and The Open University) pp. 205-230.

Strang, V. (2004), *The Meaning of Water* (Oxford: Berg).

Strang, V. (2008), 'Thematic introduction: Ownership and appropriation', in V. Strang (ed.) *Ownership and Appropriation* (Auckland: University of Auckland) pp. 6-7.

Taylor, A. (1990), *Tramore: Echoes from a Seashell* (Waterford: No publisher).

Taylor, A. (1996), *Tramore of Long Ago* (Waterford: No publisher).

Taylor, L. (1995), *Occasions of Faith: An Anthropology of Irish Catholics* (Philadelphia: University of Pennsylvania Press).

Taylor, L. (2007), 'Centre and Edge: Pilgrimage and the Moral Geography of the US/Mexico Border', *Mobilities* 2(3): 383-393.

Temple Spa (2009), *Temple Country Retreat and Spa Brochure* (Horseleap: Temple Spa).

Thesing, G. (2004), 'Revitalise yourself in West Cork', *Business and Finance Magazine*, 15 July.

Thoms and Sons (1911), *Thom's Commercial Directory* (Dublin: Thom and Co.)

Thrift, N. (2004), 'Intensities of Feeling: towards a spatial politics of affect', *Geografiska Annaler* 86: 57-78.

Thrift, N. (2008), *Non-representational Theory: Space/Politics/Affect* (London: Routledge).

Thurber, C. and Malinowski, J. (1999), 'Summer Camp as a Therapeutic Landscape', in A. Williams (ed.) *Therapeutic Landscapes: The Dynamic between Place and Wellness* (Lanham: University Press of America) pp. 53-70.

Timothy, D. J. and Conover, P. J. (2006), 'Nature religion, self-spirituality and New Age tourism. Tourism', in D. J. Timothy and D. H. Olsen (eds) *Tourism, Religion and Spiritual Journey* (Abingdon: Routledge) pp. 139-155.

Timothy, D. J. and Olsen D. H. (2006a), 'Tourism and religious journeys', in D. J. Timothy and D. H. Olsen (eds) *Tourism, Religion and Spiritual Journeys* (Abingdon: Routledge) pp. 1-21.

Timothy, D. J. and Olsen, D. H. (eds) (2006b), *Tourism, Religion and Spiritual Journeys* (Abingdon: Routledge).

Tolia-Kelly, D. (2006), 'Affect - an ethnocentrist encounter? Exploring the 'universalist' imperative of emotional/affectual geographies', *Area* 38:2, 213-217.

Towner, J. (1996), *An Historical Geography of Recreation and Tourism in the Western World, 1540-1940* (Chichester: John Wiley).

Tuan, Y.-F. (1974), *Topophilia. A study of environmental perception, attitudes and values* (New York: Columbia University Press).

Turner, V. (1973), 'The Center out There: Pilgrim's Goal', *The History of Religions* 12:3, 191-230.

Turner, V. (1977), *The Ritual Process. Structure and Anti-Structure* (Ithaca: Cornell University Press).

Turner, V. and Turner, E. (1978), *Image and Pilgrimage in Christian Culture: Anthropological Perspectives* (New York: Columbia University Press).

Twiss, H. (1928), *Mallow and Some Mallow Men* (Cork: Guy and Co).

Urry, J. (2002), *The Tourist Gaze* (London: Sage).

Valenza, J. M. (2000), *Taking the Waters in Texas: Springs, Spas and Fountains of Youth* (Austin: University of Texas Press).

van Tubergen, A. and van der Linden, S. (2002), 'A brief history of spa therapy', *Annals of the Rheumatic Diseases* 61: 273-275.

Varner, G. R. (2002), *Sacred Wells. A Study in the History, Meaning and Mythology of Holy Wells and Waters* (Baltimore: Publish America).

Vickers, A. and Heller, T. (2005), 'Traditional, folk and cultural perspectives of CAM', in T. Heller, G. Lee-Treweek, J. Katz, J. Stone and S. Spurr (eds) *Perspectives on Complementary and Alternative Medicine* (Abingdon: Routledge, Taylor & Francis and The Open University) pp. 293-324.

Voya Seaweed Baths (2009), *Voya Seaweed Baths Brochure. Organic Beauty from the Sea* (Strandhill: Voya).

Wakefield, S. and McMullan, C. (2005), 'Healing in places of decline: (re)imagining everyday landscapes in Hamilton, Ontario', *Health & Place* 11: 299-312.

Walton, J. (1983), *The English Seaside Resort: A Social History 1750-1914* (Leicester: Leicester University Press).

Walton, J. (2000), *The British Seaside* (Manchester: Manchester University Press).

Wear, A. (1992), 'Making sense of health and the environment in early modern England', in A. Wear. (ed.) *Medicine in Society. Historical Essays* (Cambridge: Cambridge University Press) pp. 119-148.

Weir, A. (1979), 'Sweat-houses and simple stone structures in County Louth and elsewhere in Ireland', *Journal of the County Louth Archaeological and Historical Society* 19:3, 185-197.

Weir, A. (1989), 'Sweathouses: Puzzling and Disappearing', *Archaeology Ireland*, 3:1, 10-13.

Weir, A. (2009), *'Irish Sweathouses and the Great Forgetting'* http://www.irishmegaliths.org.uk/sweathouses.htm/ Accessed 28 January.

Whorton, J. C. (2002), *Nature Cures: The History of Alternative Medicine in America* (Oxford: Oxford University Press).

Wightman, D. and Wall, G. (1985), 'The Spa Experience at Radium Hot Springs', *Annals of Tourism Research* 12: 393-416.

Wilde, W. (2003) [1849], *The Beauties Of The Boyne, And Its Tributary, The Blackwater* (Headfort: Kevin Duffy).

Wiles, J. and Rosenberg, M. (2001), '"Gentle caring experience': Seeking alternative health care in Canada', *Health & Place* 7: 209-244.

Wilkinson, R. (2005), *The Impact of Inequality: How to Make Sick Societies Healthier* (New York: New Press).

Williams, A. (1998), 'Therapeutic Landscapes in Holistic Medicine', *Social Science and Medicine* 46:9, 1193-1203.

Williams, A. (1999a), 'Introduction', in A. Williams (ed.) *Therapeutic Landscapes: The Dynamic between Place and Wellness* (Lanham: University Press of America) pp. 1-11.

Williams, A. (1999b), 'Place identity and therapeutic landscapes: The case of home care workers in a medically underserved area', in A. Williams (ed.) *Therapeutic Landscapes: The Dynamic between Place and Wellness* (Lanham: University Press of America) pp. 71-96.

Williams, A., (ed.) (1999c), *Therapeutic Landscapes: The Dynamic between Place and Wellness* (Lanham: University Press of America).

Williams, A. (ed.) (2007), *Therapeutic Landscapes* (Aldershot: Ashgate).

Wilson, K. (2003), 'Therapeutic landscapes and first nations peoples: An exploration of culture, health and place', *Health & Place* 9:1, 83-93.

Wilson, W. (1786), *The Post-Chaise Companion.* (Dublin: No publisher).

Wood-Martin, C. (1892), *A History of Sligo: County and Town from the close of the Revolution of 1688 to the Present Time* (Dublin: Hodges, Figgis and Co).

Wylie, J. (2007), *Landscape* (Abingdon: Routledge).

Young, A. F. (2002), *Old Portrush, Bushmills and the Causeway Coast* (Catrine: Stenlake).

Index

Page references to boxes and photos are in italic and any glossary references have g after them.